Tethered Mercury

A Pilot's Memoir: The Right Stuff...but the Wrong Sex

**Bernice Trimble Steadman
with Jody M. Clark**

AVIATION PRESS
TRAVERSE CITY, MICHIGAN

Copyright © 2001 by Bernice Trimble Steadman and Josephine M. Clark

All rights reserved. No part of this book may be reproduced or transmitted in any form or by any means, electronic or mechanical, including photocopying, recording, or by any information storage and retrieval system, without permission in writing from the publisher.

Published by Aviation Press
P.O. Box 5613
Traverse City, MI 49686-5613

Publisher's Cataloging-in-Publication Data
Steadman, Bernice Trimble.
    Tethered mercury : a pilot's memoir : the right stuff but the wrong sex / Bernice Trimble Steadman, Josephine M. Clark — Traverse City, MI: Aviation Press, 2001.
       p. ill. cm.
    ISBN 0-9709016-0-7

    1. Trimble, Bernice Steadman 2. Women air pilots — Biography. 3. United States. National Aeronautics and Space Adminstration — Biography.
    I. Clark, Josephine M. II. Title.

HD8039.A4 .S74   2001   2001-088677
629.130 / 92   —dc21   CIP

05 04 03 02 01 ✦ 5 4 3 2 1

Project Coordination by Jenkins Group, Inc. • www.bookpublishing.com

Printed in the United States of America

I dedicate this book in loving memory to my son Darryl, whose love sustained me always, and to my son Michael, whose strength I rely upon so much. Also, in loving memory to my parents, Robert Harry Trimble and Laura Christie Trimble-Whipple, whose courage has been the inspiration for my life; to my dad Ray Whipple and my brother Dick and, finally, to my husband Bob, who has been the wind beneath my wings for more than forty years.

Contents

Illustrations   ix
Special Storytellers   xiii
Timeline   xiv
Foreword by Jane Briggs Hart   xv
Acknowledgments   xvii

I Dream of Flight   3
Friendship   13
Hangar Flying   21
A Bridge of History   30
On My Own   35
Aerobatic Snapshots   47
Moving On   50
Ninety-Nine Snapshots   56
Racing: To Cuba with Joan   65
The Small Race   77
A More Complicated Air   83
I Bite the Bullet   91
Teaching to Learn and Making the Sale   104
Different Races, Different Faces   111

Hang On Janey Hart	121
Mercury Women	129
Caribbean Interlude	138
The Real Stuff	145
Tethered Mercury	157
1963 International Race	166
WACOA	172
Racing: The 1966 Transcontinental	180
Family Snapshots	189
Surgery	208
IWASM	219
Looking Backward to Shape the Future	231
Appendix A: Biographies of Mercury Women: 2000	*247*
Appendix B: Development of Women Astronauts 1960-1999	*252*
Additional Reading and Addresses	*254*
Glossary	*257*
Bibliography	*262*
Index	*265*

Illustrations

1. Mother and Ray — 5
2. Uncle Bulgie with a student — 8
3. Airman's Identification required during WWII — 11
4. Peanut Plans — 15
5. Berniece Bowers, Libby Babb and B go ice fishing — 19
6. Flint breakfast bunch, circ. 1945 — 24
7. Berniece and Bob Vaillancourt:newlyweds — 26
8. It was the Spirit of St Louis to me! — 35
9. On my way to CFI — 41
10. Flying school teachers — 42
11. The Pilot! — 49
12. Miss B with Link Trainer. — 50
13. B Steadman, International President — 60
14. Carol pours the bubbly! — 63
15. Joan Hrubec, Miss B and the Cessna 180 — 66
16. 1955 race schedule — 67
17. Designated airports — 68
18. Official flight record — 70
19. Look at all those trophies! — 72
20. Jane Hart's F-model Bonanza — 74
21. Cross Country News — 81
22. Michigan SMALL Rally 2000 — 82

23.	The Ercoupe had a simple panel	86
24.	Miss B and her Jag	89
25.	Trimble Aviation	92
26.	The new owner with her first plane	93
27.	Miss B instructing the 10th Air Force	95
28.	Miss B checks out her secretary	96
29.	Under construction	99
30.	Proud owner: Trimble Aviation	101
31.	Miss B surrounded by students	105
32.	The pilot matures: circa 1957	109
33.	Lucille Quamby, Betty Hutton . . . Miss B	113
34.	Mary Clark races for time stamp	117
35.	The E6B	118
36.	Janey and B	122
37.	Trimble Aviation letter	134
38.	Albuquerque schedule	135
39.	The *Harebell* exterior	139
40.	The *Harebell* interior	140
41.	I was invited!	144
42.	Jerri Sloan	145
43.	The only graph I ever saw	147
44.	Cochran letter, July 12, 1961	159
45.	Cochran letter, August 1, 1962	165
46.	Mary and B admire their 1963 winnings	170
47.	Janey's telegram	173
48.	Letter from Phil Hart	174
49.	Airman's Subcommittee submission	175
50.	WACOA circa 1964	179
51.	The Comanche panel	184
52.	Best in Class	186
53.	The Winner!	187

54.	AWTAR participation map	188
55.	Pensive little miss	191
56.	Pretty as a picture	194
57.	Bob looks great in law duds, too	195
58.	Carol, Aunt B and Elizabeth	197
59.	The Muscleman in school	198
60.	A busy boy and his elephant stick	200
61.	Darryl and his big brother, Mike	206
62.	Chef Michael Steadman	207
63.	I'm a survivor!	209
64.	Medical journal clipping	217
65.	Page Shamburger	219
66.	IWASM logo	221
67.	Nancy celebrates her birthday	225
68.	We visit our spot	245

Special Storytellers

Page	Speaker
13	Berniece (Bowers) Vaillancourt
21	Berniece and Bob Vaillancourt
56	Berniece (Bowers) Vaillancourt
65	Joan Hrubec
83	Bob and Berniece Vaillancourt
121	Janey Hart
145	Janey Hart
157	Janey Hart
172	Janey Hart
189	Bob Steadman Denny Whipple Mike Steadman
208	Bob Steadman Berniece (Bowers) Vaillancourt

Timeline

1925	Born Rudyard, Michigan
1932	Mother remarries
1943	Meets Berniece Bowers
1944	Flying lessons start
1945	Earns private license
1946	Begins to work for Francis Aviation
1949	Begins to work for Flint Flying Service
1956	Manager, Flint Aeronautical
1957	Opens Trimble Aviation
1959	Marries Bob Steadman
1961	Trip to Caribbean
1961	Mercury Astronaut physical
1964-1968	WACOA member
1967	Steadmans adopt Darryl
1968	Sells Trimble Aviation
1968	Ann Arbor Airport Commissioner
1968	Works for Twining Aviation
1969	Cofounder IWASM
1972	Steadmans adopt Michael
1974	Opens B Steadman & Company
1975	Neurological Surgery
1977	Works for VISTA

Foreword

Janey B. Hart

Tethered Mercury is the story of Bernice Trimble Steadman, who was born into a family of the most ordinary means in the Upper Peninsula of Michigan. Tragedy struck one night when their house burned down in a fire, which cost the life of her father as he tried to rescue her other siblings. Her mother took B and left the U.P. for Pontiac, in the Lower Peninsula, where she had some family.

Bernice was a high school senior when she decided to become a pilot. She had no money for such a project. Her stepfather worked at the Chevrolet plant in Flint as an electrician. Bernice applied for work in the research department at A. C. Spark Plug because, after all, she was a high school graduate with good grades in math. But she lacked a degree in engineering so she became a spark plug inspector. Self-confidence *is* an important character trait in a pilot, melded with a healthy dose of caution.

WW II was in progress when B went to work, so there were a lot of women doing plant work with her. Her paychecks were barely enough to pay for thirty-minute flying lessons three times a week. She worked hard and studied hard: aerodynamics, engines, meteorology and navigation to pass the written test. She finally got her private pilot's license before she got a driver's license.

Step by step B upgraded her skills and ratings. I was one of her students, thank heaven, because when I took my instrument flight test the inspector wrung me out for two and a half hours through every approach, with an engine out every time. Since those exciting days, we have enjoyed a solid, wonderful friendship. The older I get the more I realize that friendship is the most valuable relationship in life.

So now, fasten your seat belt. You are in for a good American story.

Acknowledgments

The decision to write my memoirs did not come suddenly. It was suggested by many friends over the years and finally, when it seemed necessary to explore fully the story of the Mercury 13, my husband Bob and I began putting events down on paper.

Writing this book has turned out to be an interesting adventure. As Kahlil Gibran wrote in *The Prophet*, "A friend is your needs answered." By his and any other definition, I have truly gained a friend in Jody Clark. She agreed to help and it is her collaboration that made the project possible to complete. Her writing and organizational skills are superb. I also value the tremendous support given by various people over the years, who made my story rich in adventure. I introduce you to many of them in the book, but it is not possible to include them all.

My family has been uniformly supportive, as have Janey Hart and Berniece and Bob Vaillancourt. As one life is but the sum of many, in the book they share some personal memories with you, the reader. In the process, I hope they have enjoyed revisiting past times and events as much as I have.

Although the opportunity to serve in the space program was never given, I count my friendships with the other Mercury Women as one of the great blessings in my life. They were, and are, vibrant, intelligent and motivated. They have dealt with the devastating disappointment of the end of this opportunity in different ways, but always with humor and a respect for each other. We missed going into space as Mercury Astronauts, but gained a tremendously enriching experience because of our friendships. It is largely because of their support and the public interest in the question of women in NASA's Mercury Program that I have yielded to requests to write about my life in aviation. You will not actually hear the voices of some of my friends in this book, but in your mind's eye I hope you can see their rich zest for life and their love of aviation as they share a few of their adventures.

I will be enormously gratified if you enjoy flying through these pages with me. Life in aviation has never been boring.

BERNICE TRIMBLE STEADMAN

My collaboration on the story of the Mercury Women has been fun. I've learned a lot of history and spent time with fascinating individuals, women who made choices I didn't make, but that I admire. They are active people who forge strong bonds of friendship. Some of them might not have had contact with B for years, but when details were needed for the book, the material was rapidly made available. Sometimes events pull people in different directions, but their support of each other's projects holds fast and reinforces the bond.

Tethered Mercury is based on a series of stories told by B. Steadman and some of her friends during the late 1990s. It represents material they remember from the time they were young into the present (October 2000). While every attempt was taken to document the stories for accuracy, the reader needs to remain aware that these stories are memories, a record in the vocal tradition, lovingly told, reaching through time across several generations.

We would like to thank everyone who played an important role in creating this book. In addition to the people you will meet as you read the story of Miss B, Dawn Buchanan, Amy Carmein, Pat Ponczek and Martha Vreeland gave generous, tremendous help. The Grand Traverse Writers Group also left their mark on improvements. Mike Stock, Chief Flight Instructor for Northwestern Michigan College's Aviation Department, improved the glossary and brought it up to date. Finally, with a sense of relief, we depended on Susan and Nikki of the Jenkins Group to fly the book through design and printing.

In eight years of steady work, many stories were told and many were selected for you to enjoy. Does the end of the book mean the end of the stories about the Mercury Women? Not in a million light years of traveling on Spaceship Earth. If you meet one of these women, ask her to tell you a story. They each have a fascinating tale to tell. I have helped sketch the life of just one of them and she isn't done yet. For an example, a new story arrived last week, "My mechanic, Perry Simons, thinks we need to include this story about meeting Werhner von Braun. Can we add it please?" I agreed.

JMC

Tethered Mercury

I Dream of Flight

I cannot remember not wanting to fly. Although it was more than fifty years ago, when that incredible day pops up in my memory, I still get goose bumps. The morning sun was bright, a bit of dew sparkled on the grass and small puffs of white cumulus slowly drifted through a clear blue sky. The perfect day for a perfect event, my first flying lesson. At seventeen, the idea of flying lessons felt like the culmination of my whole life. My stomach was full of butterflies.

First Flight

My desire to fly started long before Clarence Chamberlain flew into Flint, Michigan, with a Curtis Condor in 1937. The newspaper said that he was offering rides and I knew from the moment I read the article, that I just had to go for a ride with him. At twelve, I was an avid reader of library books and had read about Chamberlain's flight across the Atlantic shortly after Lindbergh's. I knew my brother Dick could always twist Dad around his little finger. I figured the only way I could get a ride in the big, black, magical airplane was to get him to want to go so he would ask our parents for the money to purchase the tickets. I promised to let him sit in the window seat and told him about the wonderful time we would have. The plan worked. Money in hand, we raced to the plane for our turn to fly.

Ushered in from the rear, we walked between wicker seats up a narrow aisle to the front of the plane, where we sat right behind the pilot. Dick settled in at the window with his back flat against the chair. When Chamberlain took off, my golden-haired brother glued himself tighter to his seat and kept his eyes straight ahead. I leaned over him to look out the window and, for the first time, felt the fascination of watching the earth pass below my wings.

A .C. Spark Plug

I grew up in times of little money. In 1945 my dream of going to college to become a doctor was met by Dad's reply, "Spend a year working and learn the value of a dollar." The disappointment of not going on with formal education intensified what became a burning desire to fly. While finishing my senior year of high school, I spent the winter dreaming my way through a selection of advertisements for flight schools located all around the country. Eventually I realized the money required to go away would be just as impossible for me as medical school, so I called a local flight school for an introductory lesson. But I still needed money.

With the war not yet over, jobs were available in defense plants. A. C. Spark Plug (AC), a division of General Motors (GM), had a plant in Flint making spark plugs for both military and automotive use. A graduation certificate from my school indicating courses successfully completed in math and science gave me priority for a job in the defense area. Armed with this certificate, I marched down to AC to tell them I wanted to be in their experimental department. Given the circumstances, they were kind to me. It was probably the first time any high school graduate had applied for a job requiring a college engineering degree.

The job they offered me was touted as something special although it sure wasn't my idea of research. Inspecting spark plugs also didn't seem to be very important, but I couldn't very well turn the job down. On the plus side, I soon figured out that it would pay enough to fund flight lessons, with a little left over. I remained confident I would be part of AC's experimental department in a few days. Looking back, I have laughed a lot about my presumption and cannot imagine having been so self-confident.

Most of the people I worked with then at AC were women whose husbands or sweethearts were in the service. Others wanted to feel they were doing their part in the war effort. I just wanted to fly. They all treated me with respect for my dream and the job was tolerable because of their friendship and support.

Mother

I continued living at home. Mother said she would not stand in the way of my flying but would not help finance such a crazy idea. She had seen too many movies and heard too many stories about pilots' devil-may-care attitude and penchant for drink. I must admit that devil-may-care reputation was part of the fascination for me but the drinking was an unknown quantity at the time.

My mother was a great homemaker, really a professional homemaker. No one came to our house without smelling some-thing great cooking.

Mother and Ray

When we were little kids and one of my friends announced she was running away from home, Mother packed up some cookies and I joined the friend who was leaving. I wasn't going to leave home, my friend was. But, I couldn't let her go by herself, so we'd head off with our bag of cookies way down the railroad tracks. We would tell our mothers we were going to hop on the freight train for some wonderful destination. We used that threat to scare our mothers. Then, as we walked, it would begin to scare us. Still, we walked down the tracks, swinging our bag of cookies, talking about all the things we were going to do and the places we would go when we grew up. Mother never said don't go; she just always made sure we had cookies. What a wise mom. She seemed to know little girls had to have a time to talk and grow up. The cookies just made it all seem sweeter. Of course, when the cookies were gone, we headed for home.

FIRST FLYING LESSON

Now that I was secure in my job, on that first spring day, I pedaled across town with my mouth as dry as a bone and my mind full of stories about earlier ladies who had taken to the air. I'd read everything I could ever get my hands on about them: Amelia Earhart, whose daring acts of courage were as well known as the flying feats of Charles and Anne Lindbergh (perhaps I would blaze a new trail from north to south); Phoebe Omlie; Teddy Kenyon; and others who were test pilots for new aircraft or specialized instruments which would allow all-weather flying (for now, the thought of these women satisfied the engineer in me). Then there was Blanche Noyes. She had the responsibility of air-marking towns to aid pilots navigating by dead reckoning, which means flying by ground reference only. Each of these pioneers conjured up an excitement I knew I wanted to share, if only I could.

Turning onto Torrey Road meant the last mile or two would be on gravel, my easy ride over the pavement ended. As the road curved, it rose up to cross railroad tracks, the highest point in the trip. Suddenly I could see the control tower and runways of what I hoped would become my second home. As I watched airplanes landing and taking off, exhilaration fueled the butterflies in my stomach and I had great difficulty pedaling that last mile to my destination.

Finally, willing my bike to the airport, I introduced myself to the woman sitting behind a misused desk in the office of Francis Aviation. Middle-aged, Jean Ramsey was anything but the dashing young man I had expected to see. She was, however, very much in control. In no time, I was under her spell.

"Just why are you here and why do you want to learn to fly? It's expensive. We expect students to fly at least twice a week. Otherwise they forget too much and it takes too long to teach them to fly safely."

Safely! I surely wanted to fly safely. But the twice a week requirement meant I could only fly a half-hour at a time because that's all the money I had. Still, this did not deter me. By now I was ready to leave the earth behind for any amount of time. As Jean grilled me, I began wondering about my instructor. What credentials would he have? I expected him to have vast amounts of experience, but wondered if he would be patient and forgiving of my awkward beginnings. Could I let him know how very much I wanted to fly without his laughing at me? More butterflies!

Satisfied with my sincerity, or giving up trying to discourage me, Jean finally introduced me to Burr Walton. I did my best to appear confident as we walked to the plane and began the preflight routine: checking fuel, oil and tires. While we made sure the controls moved freely, Burr identified each of them and told me what each did when the airplane was in flight. It all sounded professional and complicated, but I drank it all in. I think he could see from my expression that it was time to take this fledgling up to see if she would become part of his flock.

When we were finally in the air, I was so excited that I watched the motion of the wheel (it was not a yoke) and rudders for some time without even looking out the window. Burr put the little plane through some gentle turns, climbs and glides. Finally he turned to me. "Put both feet on the rudder pedals. Lightly now, put one hand on the wheel and follow me through on the controls."

After feeling the motion of the controls so I could sense what the ailerons, rudder and elevator would do, Burr finally allowed me to take full control. Further instructions came intermittently, but loud and clear. My first job was to try to make the Taylorcraft fly in a straight line. "Just follow that road and try to stay at two thousand feet." It sounded like a simple request, but it took time to master. Burr was in the right seat and I was in the left, or pilot's seat, as we flew. Our T-craft was light, powered with a sixty-five horsepower Lycoming engine and had long, thin wings. As a result of this high aspect-ratio, I soon learned this plane would take off after a short run, but was a bit tricky to land.

Burr gave me preliminary lessons two or three times. He was a good instructor, but moved on to bigger and better things, leaving me behind with an all-consuming, fierce desire to continue. Even that first time, when it was my turn to take the controls, all the butterflies in my stomach just disappeared. If my stomach was comfortable, this had to be right for me!

Ralph Rose (Uncle Bulgie)

My second instructor was a colorful, short man with a ruddy complexion that told of his long appreciation of Demon Booze. Were all my mother's fears to be right? Should I trust this stocky man with my life? I gradually learned that Ralph Rose, fondly nicknamed "Uncle Bulgie", had been flying for years. First, he entered the Army Air Corps with a great desire to participate in the war as something other than a foot soldier. He also wanted to get out of the war with his whole skin and an officer's rank. Considering what he wanted, it is ironic that he wound up flying troop gliders into battle areas, a mission considered extraordinarily dangerous for both the troops and the pilots.

Through these stories and more, I pieced together his history: When he finally was released to enter the civilian flight game, he was older than most of the other pilots looking for airline jobs. Ralph was a bright, self-educated man who seemed to have read everything written about aviation in fiction and non-fiction. He decided to apply for a bush pilot job in Alaska. Since he had no experience flying in the bush, he reviewed the conditions confronting the Alaskan pilot and realized survival in the Arctic was not easy for the novice. Undaunted and sure of his ability,

Uncle Bulgie with a student.

north he went with the cockiness of a new cadet and the assurance of a high-time, mature military pilot. In Alaska, he quickly learned on-the-job instrument flying. Landings on remote lakes found this clear-thinking pilot with more than a normal desire to survive. With repeated success, Ralph was welcomed as a member of the inner circle of Bush Pilots. However, whiteouts were a too-frequent fact of life and arrived with little warning, so when his contract ran out, he returned to the lower forty-eight. As a student at Francis Aviation, listening to his stories, the fact that he managed to come out alive seemed a miracle to me.

Uncle Bulgie soon had me ready for the next step, learning to land the plane. Mastering straight and level flight high above the ground with Burr Walton faded into an easy memory as I struggled to slow the T-craft enough to feel my way back onto the runway. There is a point within a few feet of the runway where a pilot starts to ease back on the elevator to flair. Done right, this flair brings the plane to a stall just at the moment it touches the runway. In the Taylorcraft, it is tricky to neither run out of lift too soon, nor fly off the end of the runway. I had some uneasy moments while I learned to control my plane, which was really little more than a powered glider.

Struggling to Fly

Even in half-hour increments, flying twice a week was a challenge since aircraft flight rentals cost $10.00 per hour, including the instructor. The meager salary I made inspecting spark plugs netted me a take-home pay of $27.00 a week, so I walked to work. After a while I learned I could purchase a block of ten hours of flight time for $64.00. Even with this bargain, the requested twice a week lessons meant I continued to fly for less than an hour a lesson. I did not own a car and had no license to drive even if I'd owned one. With still-limited finances, a bicycle often had to do for transportation.

On Saturday and Sunday I biked to the airport to fly for a half-hour. Sometimes I took a bus, but then I had to leave early enough to make the necessary transfer downtown to a second bus that went to the outskirts where the airport was. Both the bus and the bike trip took an hour and a half to two hours. The length depended on the weather for either alternative. Sometimes the trip was an excursion in time, topped with great exhilaration, and sometimes it was pure frustration. Regardless of choice, I usually arrived at the airport frozen or fried.

My desire to fly was overwhelming. I'd call the airport before I left home if the weather was already bad. Ralph always said to take a chance on the weather clearing and come on out. If I got there and we couldn't fly we would play a form of poker he called "red dog". In this game, he could change the rules to win, thus claiming I had to buy his coffee or, worse, his lunch. My best friend, Berniece Bowers, had a fit when we did this because I was gambling my hard-earned money, but I didn't mind. Uncle Bulgie was a marvelous storyteller. I could sit and listen to him for hours. After all, he was talking flying!

Solo

After nine hours of flying in bits of time that vanished far too rapidly, I arrived at the airport one windy, cloudy day for an expected lesson of takeoffs and landings. Ralph was obviously suffering from the effects of a night out with the boys, in no condition to take many bad landings, or even to have to explain much. After several of my attempts at a perfect landing, he motioned me to turn off on the grassy north end of the runway. "If you're going to fly like that, you're going to have to do it by yourself!" Without so much as a "You take her around," he climbed out of the plane and walked away. "Oh, my god! He's so scared he's got to get out!" There I sat. ALONE!

All my surface confidence flew out the door when Ralph left. I don't really know how long I sat there, but finally a new worry intruded. The only thing worse to me than not taking off would be to face everyone without attempting the solo.

I began to act on Ralph's challenge. "Taxi onto the runway. Poise for takeoff. Glance at the tower for a green-light clearance." In 1945 we had no radio. They used light guns for communication. Green was "go" and red was "don't go". Alternating red and green was "caution:" look around for traffic on the ground or in the air. I drew a deep breath.

"With a firm hand, ease the throttle in." Obedient to the routine, the plane began to accelerate down the runway and, before I could catch my breath, I was airborne. Alone, the cockpit seemed very empty, but the ship felt light and maneuverable without Uncle Bulgie's weight. The extreme feeling of freedom was overwhelming; the air was truly my domain.

Then, suddenly, I realized the increased maneuverability only meant that I would have to get back to the earth on my own. My heart began to pound so hard it drowned out the quiet instruction from my subconscious mind. Then, quickly, words, all the words I'd been listening to in half-hour bits for so many weeks, seemed to become audible. "Fly straight ahead to an altitude of four hundred feet, check for traffic, make left turn, ninety degrees. Level off. Fly straight and level, then climb to six hundred feet, check for traffic, left turn, ninety degrees. Level off. On downwind climb to a pattern altitude of eight hundred feet. Fly straight and level. Begin to throttle back as you turn ninety degrees to base. Check for incoming traffic. Pull carb heat. Turn to final approach to land into the wind. Just above the ground, slowly pull the nose up to flare to land."

Coming around the airport traffic pattern to set up the position for my first landing, a glance at the tower showed the red that indicated a go-around. I couldn't see any reason for it, but as instructed by the light, I added power and climbed back into the pattern. Coming around again, I remembered a story Ralph had been telling his devoted, mesmerized students. He really didn't think much of women flying, but with the war on, the gullible women who listened to him over coffee were his bread and butter. We believed every word he said. He was so revered by us, and by the few men trying to build up enough time to get into Army Air Corps enlistment, we were sure he could walk on water. Still, he was not above a good story meant to keep this flying sport a man's world. He claimed that a woman he soloed was so frightened that she froze and would not come down to land. We waited around the table until Ralph finally took the cigarette out of his mouth to deliver the punch line. "I got out my shotgun and shot a hole in the gas tank. When she ran out of fuel, she landed." Well, I wasn't going to let the people in the tower, or

Ralph, create a silly new female story. When I got the green light to land, I flew the plane onto the ground with a student-perfect touchdown. Then I continued the required three landings and returned to the hangar one mighty happy gal. I was alive!

It was customary upon solo, and I might add in Ralph's case it was a requirement, to present your instructor with a bottle of whiskey. Since these were the war years, whiskey was a rationed item, hard to obtain. But obtain it we did. Mine arrived via truck from some unknown source never to be revealed.

Pilot!

As I was flying, I was also studying the mechanical systems of the Taylorcraft. To pass both the written and flight examinations for my license, I wanted to show a professional familiarity with the physical aspects of my plane. After I flew the minimum time and the cross-country flights required for a private pilot license, Ralph told me to prepare the necessary papers for my Flight Check. The first butterflies in a long time instantly returned. My image of flight examiners was that they were tough to please. If they didn't give me a rough time, they would be banned from aviation.

Airman's identification required during World War II.

I also felt that this was my day of reckoning for a second reason. Now I had to tell Ralph that I didn't have a driver's license. As embarrassing as this was to me to have to admit, it was absolutely hilarious to him. He could not get over the fact that in all this time I had never mentioned not owning a car.

Once he got over his fit of mirth, he set in motion what seemed to me to be the most hair-raising part of my entire flight program. Jean was told to drive me home to get my parent's consent on my application and to be quick about it. The Inspector was due any minute. Driving like a crazy woman, Jean convinced me the only way to travel was by air. Never have I had such a ride, not even when I later earned a reputation for fast cars. My mother reluctantly signed my consent form, Jean reversed the process and the rest became history.

Friendship

(with Berniece Bowers Vaillancourt)

As time went on the young women who came out to Bishop Airport to learn to fly included my best friend, Berniece Bowers. Berniece is bright and witty, has an impish voice, a merry sparkle in her eye, and can be hard to keep up with.

(Berniece) I could see I had to fly to remain with my friend. When the new girl moved in across the street from me with her mother, father and brother, she was fourteen or fifteen. I remember I didn't invite her to my sixteenth birthday party because I was angry with her about something. She was a new phenomenon in my life and it took a while to adjust to the new neighbor who was also called by my name.

We moved a lot once my mother remarried. I used to say we moved every time the rent came due. Mother didn't like that very well, but I was always moving out of one school district into another. My brother, who was seven years younger, didn't have to change schools because we always moved around in the same elementary district. I felt I could complain. Once we got over on Ohio Street near Berniece, we finally stayed in one place.

You moved into an identical two-story house directly across the street from me. It was a middle-class blue-collar neighborhood, with just about everyone working at the General Motors plants.

My dad was a factory worker; Berniece's dad was a factory worker. Her dad was on supervision and mine was an electrician. Ohio Street was considered to be part of a nice neighborhood, where factory people bought or built houses and stayed in them. There were no rentals.

My new friend Berniece and I walked to and from school together, talking back and forth. She'd ask me to hold her books for a minute and I'd wind up at school or at home still rattling on, still carrying her books.

It wasn't an immediate thing, but after we became friends, we did a lot of church things together. I made sure she got into the choir, Girls Friendly and the young people's group. Girls Friendly was like Girl Scouts, but in the Episcopalian Church. I'd laugh at the boys when they'd ask if I was a Girl Scout. "No, I'm a Girls Friendly." We took Bible class together, played a lot of softball, went swimming and would ride our bikes five miles to go horseback riding once a week. We also built model airplanes and showed them off to each other.

Model Airplanes

I got interested in building models by watching my cousin, Leonard Jacobson, who lived in Sault Sainte Marie. We traveled north once a year to be with my mother's large family. The Christies gave me a window to the past and surrounded me with a feeling of love and continuity. In the early summer, as we picked wild strawberries, we watched in fascination as the new, shiny DC-3's landed and took off at the airport.

While my cousin built his models in our free time, I could only watch, not touch. I would stand behind him for days, out of the light, as he shaped small, thin pieces of balsa wood with a razor blade, fitted them together, and covered them with thin paper. A bit of spray tightened the tissue and each would finally become a marvelous flying model. This happened for two or three years before I decided I was going to try to build one. The darn little model didn't know I was a girl! One model. It went together and flew beautifully. That was all it took to set me on a different course.

Just like Leonard, Berniece and I made our airplanes out of thin balsa wood covered with Chinese tissue paper. Common pins held the sheet of balsa flat. We taped the plan for the plane we wanted to build over this balsa, cut the stringers out to form the shape of the fuselage, cut ribs to form the shape of the wings, applied tiny dabs of glue, and held a thing of beauty in our hands. The finished planes were rubberband-powered and a wonder to fly. But, once one was finished, I had a hard time keeping it out of my little brother's reach.

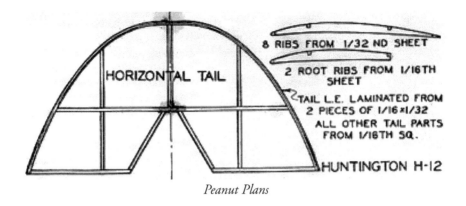

Peanut Plans

B's brother was really jealous of her efforts. She would just get one built and he would use some excuse to smash it. He was too young to build them. I would complain to her beautiful mother, but Laura could hardly cope with this young rascal.

After Berniece and I actually started flying, Cousin Leonard decided to learn how to fly and got airsick every time he went up!

"Pow! We got him!" We loved it when we finally triumphed over Leonard because he felt so superior to us, even when our models were successful. Leonard always threw all sorts of airplane terminology at us. We were reading flying magazines to keep up with him. We also read all the novels with any aviation basis in order to talk to the men at the airport. If you talked about a P-51 with them, you had to know what horsepower it had. I never really understood very much about what I said. I just memorized stuff.

Unlike my friend, I was still in high school the first year of my flying lessons. Out at the airport, the mechanics were already calling my friend "Miss B". Determined to join her, I worked in the school secretarial program on Saturday and after school, making ten dollars a week. Seven of that went to flying. The last three dollars had to cover all the other things a girl might want. My mother would check my pocketbook and add money when I ran out, but she wouldn't talk about it. She wanted me to feel I was on my own.

Mother spent all the time I flew on her knees, praying. She would never go out to the airport, never looked at an airplane. She prayed all the time, but she trusted B. Once we decided we'd fly over the neighborhood and make sure *our mothers saw we could fly. With the sound of two planes overhead, they'd have to come out and look. I flew down very low and pulled the airplane up in a wingover. B did the same thing. I lost track of B's position. Suddenly, I was looking right at her, framed by my windshield. Without radio, it was too easy to lose contact with exactly where the other plane was located. We didn't do that ever again.*

The Four Musketeers

As soon as the group hanging out at the airport had become well acquainted, some real fun began. Our usual Saturday morning rendezvous found four high-spirited young ladies meeting to take turns playing follow-the-leader. We would take off in trailing formation, then formed into a line so the first pilot could lead us through loops, climbs and dives, trying to lose her flock.

One morning Christine was to be the leader. Libbie, Berniece and I were rather conservative pilots. We knew Chris took more chances than the rest of us did. She sometimes put us on the edge of what was almost too scary. This was actually not a totally undesirable condition. Strapping myself into the Taylorcraft, I viewed the weather with some doubts, but figured our friend would not take chances with the weather and would watch out for us so we could return to the field if necessary. With pre-flight completed, I called, "Ready! All clear! Brakes! Contact!" The ramp attendant pulled the propeller through, the engine fired and I began taxiing out for takeoff in the middle of a four-plane formation.

Shortly after we were all off the ground, Christine started an abrupt climb. We followed closely. When she leveled off, we were almost in the cloud bases, with poor visibility. I was about to break out of formation when Chris dove to a lower altitude. With a quick push on the wheel, I followed my leader with great glee. The next thing I knew, the four planes were flying in a loop! I was almost at the top of the loop before I realized what Chris had gotten us into. With four aircraft going up and around, I suddenly looked out the windows and could not see my wing tips. When I pulled out at the bottom of the loop, no other plane was in sight. Panic!

My first thought was that everyone else had run into each other in the clouds. As fast as I could, I reduced power and turned for the airport, hardly daring to think of possible conclusions to this crazy game. Safe on the ground, I found the rest gathered over coffee where we soon shared our tales: as we entered the clouds, each of us had changed heading to make sure she avoided getting in the way of the others. That made us feel pretty good until we realized some of us could have taken the same evasive heading. There, in the small coffeeshop, four aspiring pilots grew up. We had been lucky, but ignorance would not be our protector for long. It was time to think of the future, set some ground rules and find out what more experienced pilots would use for evasive maneuvers.

American Aces

We called ourselves the American Aces then. We thought we were getting good enough to fly like our heroes. A friend in the country had a house we used to buzz. Once when we returned to the airport, my boyfriend told me I'd come pretty close to some telephone wires. "What wires? I didn't see them." I had flown under them! We quit that then too. We'd strafe water towers to read the name of the town when we needed to determine where we were. Once I was reported to the authorities for flying so low. "What do you want me to do, land for directions in some farmer's field?"

Cross-country flights were a requirement to get a license. They also became a group challenge. We would prepare together for these flights a week ahead, flying most often to grass strips within a hundred miles of home. Even when they were to be close to home base, preplanning for these flights was serious practice for the written exams. We would draw a flight line for each leg of the trip on the chart to plot the course ahead of time. We understood wind drift and triangulation, so we planned each leg to cross section lines at a particular angle. We paid particular attention to railroads and highways. They were good checkpoints. We also computed a rough idea of the time each leg should take to fly between checkpoints so we would not fly too far off our intended route before realizing something had changed from our preplanning.

The day of the flight, we'd arrive early to check the new weather charts and the Teletype. This gave us a running history of a number of hours of actual weather changes, predicted future weather and direction of the winds on each planned leg, listed current observations and forecasts for enroute and destination airports and, occasionally, included actual enroute pilot reports (PIREPS). As we finished the computations, we each would check out a plane and head up into the Michigan sky. Keeping track of our direction enroute was hard because the old compasses still jiggled so much. It was best to follow landmarks on the ground pretty closely. Precision dead-reckoning was not possible until they developed directional gyros to provide more reliable position information.

The state of Michigan was surveyed north-south, east-west into townships which were then divided into sections. As we flew across each section line we checked our angle to determine if we had the plane turned enough into the wind to remain on course. If the wind speed or direction changed from what we expected, we could tell by the time it took to fly the leg and the difference in our approach angle. If it wasn't what we expected, we would figure a better angle to continue to our destination and revise the estimated time of arrival (ETA).

We always planned a rendezvous at some airport along the route for a cup of coffee. This often meant waiting on the ground for some of the gang to catch up. One would have trouble starting an engine, another would start off in the wrong direction and, although all the T'craft were alike, they did not have the same cruising speed.

One day B said to follow her. When I got in the air, my seat was not adjusted correctly. I thought there was something terribly wrong with my airplane. I could hardly reach the rudder pedals. I also should have used a pillow to see over the cowling. B waited a long time for me in Lapeer, then had lunch. I had returned to Flint. I felt I was lucky to get back on land at our home base.

I used to get lost on my own too. One day I started out for Lansing and landed in Owosso. Wandering around a while, I finally went up to some people and said, "It's nice to be here. Where am I?"

Another time B and I flew to Alpena on a longer cross-country. After we arrived, we got weathered in. At the hotel, we put the key in the door and it broke off. When I went downstairs for another key, the woman at the desk got out a cigar box and told me to take any key. Well, which one was for our room? "One key fits all." A wildcat hunters' convention was in residence then, too. My mother wanted me to come right home!

An Airshow Ace

B got me into more problems and involvements! Once she challenged me to enter an airshow put on by the Flint Civil Air Patrol (C.A.P.). That put Mother on her knees for a week. B said she couldn't participate. She felt she was a professional because she'd begun working as a clerk, and was sometimes flying for Francis Aviation. In her opinion, however, it was the perfect thing for me to do. So I entered the show. I competed in spins, power-off landings and spot landings on her recommendation.

I did a lot of growing up during that show. There were five thousand spectators to watch my daring feats! I volunteered to take parachutists up. In those days, this was very risky. It was easy to destroy the balance of the forces that kept our fragile airplanes high in the air and to take an unwanted tumble. At another point in the competition, I was going to be short of the field, so I dove at the fence and just barely peeled over it.

During the bomb-dropping contest a WWI pilot sidled up to me and said I should make the drop a certain way, never mind the winddrift. But when each big old sack of lime fell, the results were terrible. My paper sacks were falling off the airfield into the parking lot. I looked down and saw people running. People were climbing fences! I thought, "Oh, God, I can't land because they'll arrest me. ...Oh well, maybe they'll forget which plane it was." When the show was written up in the newspaper: "One pilot missed the field completely." We women pilots appreciated that they didn't say it was a woman. I was glad they didn't use my name, or say the bags landed on a car. The C.A.P. wanted me to go on to Lansing for another competition, but I wouldn't. It cost too much money to rent an airplane to fly for a show.

One day I got my Dad up in the air. One day! We were in a terrible airplane. It leaked oil, had rags stuffed in the windshield... ." What kind of a crate you flying you have to hold the door shut?" "Just hold the door shut, Dad!" I loved aerobatics, so I did everything but get him nauseated. He wasn't impressed, but I had fun.

(l-r): Berniece Bowers, Libby Babb, and B go ice fishing.

I had a private pilot's license and my father wouldn't even let me drive his dumb cars! It was humiliating. If I went to cash my check, I'd show my pilot's license. In order to avoid the bus ride, I told everybody that if they would drive me to the airport, I'd fly them anywhere. I never let Dad find out about that either.

By the time Berniece flew for the C.A.P., I wanted more than anything to become the best pilot in the country. So I worked hard professionally and learned how to make all the sensitive machinery work to my command. But first, we had more lessons to learn as a group.

Hangar Flying

(with Berniece and Bob Vaillancourt)

Berniece had no taste in men. It was awful.

(Berniece) Not only did I try to follow B into aviation, I even allowed her to pick out my boyfriends for me. My taste in men? Hers didn't seem any better! But finally, there was one lineboy, Bob Vaillancourt, who B challenged me to charm, but it wasn't easy.

It also wasn't hard for me to mess up. One winter when I was flying on skis I went to the pump where some old boy put the gas in my plane. I revved up to turn away, thinking I was going to put my brakes on...but there are no brakes with skis! The gas jockey dove under the wing, fortunately, for I would have decapitated him. When I saw there was nothing wrong, I took off and flew for an hour. On my return, they said the airport manager wanted to speak to me in his office. He sternly told me I almost took the gas pump and the poor old boy out with me. I asked for a Kleenex, "please," and sat there to cry for a few minutes. The manager ended up giving me a hug, telling me he knew I wouldn't do anything like that again. I said, "No, sir, I won't. That was scary!"

Berniece and I flew quite a bit in T'crafts on skis. They didn't plow the runways as much then; they'd just roll them. In the winter we landed in a furrow with little banks on both sides. The problem wasn't that we couldn't stop. We just couldn't stop in a hurry. If we slowed the darn plane down, the skis would freeze. Then somebody would have to come out to pump the wing up and down to break us loose again. If we were out there alone, we'd have to get out of the plane, grab the handle on the tail, stick the tail into the snowbank on the runway edge, shake the wings hard until the skis broke loose, then run like the dickens to get back into the plane before the skis refroze onto the snow. I don't know how many times we froze our hands working on those planes. In addition to the complications of freezing tight to the runway, either our tailskid or our narrow, hard rubber tailwheel cut into the snow and caused drag all the time. That was something to contend with taking off *and* landing.

THE LINE BOY

(Bob Vaillancourt) B, Berniece and I all arrived at the airport about the same year. I was a line boy, starting in 1943. I remember marching around the field and having a good time with the C.A.P.

Berniece and I weren't in that group because we didn't want to march. We used to watch them try to stay in formation and laugh at them. I thought it would be awfully nice to wear their uniform though. The closest I got was when we all wore surplus flight jackets. Early on, they weren't even leather. This was near the end of the war. We were all too young to really participate. They took young men into WWII from our senior class as early as age seventeen; in the navy at seventeen and the army at eighteen. There were few boys around for our senior prom. Bob, how long were you in the C.A.P.?

I moved out of the Civil Air Patrol when I signed up with the Air Force Cadets on my seventeenth birthday, as a junior in high school. I went into the Cadets on a voluntary basis, feeling I would get a better chance at pilot training. My reason for working at Bishop was to have a bit of background in aviation to help me get a pilot's slot in the Cadets. Right after graduation, they took me into the service. but by the time I joined, the air service had pretty well wound down. My Air Force training lasted about a year, then I was out.

At Bishop Airport, when I was sixteen, I used to earn two hours of flying by working seven days a week. I'd go to the airport after school to work, plus working Saturdays and Sundays. I had to take two bus transfers, and then walk about two miles from the end of the bus line to get to the field. I saw Berniece on the way out all the time. Sometimes we met downtown, but mostly I ran into her at the end of the bus line, so we walked together from there. It wasn't until much later that I learned she planned to meet me all the time.

As a line boy, I hand-propped Aeroncas, Luscombes, Taylorcrafts. There was no such thing as an electrical system in those days. Everything was run with the old magnetos (mags). A magneto uses pulsing magnets to generate electricity to run the engine. Just about everything had to be hand-cranked, to get the mags to pulsate fast enough to carry a consistent current. I should say the pilots needed somebody to do the cranking. I was that somebody.

Since the mags created only enough electricity to keep the engine going, we had no radio communication. We flew following the guidance of flashing lights from the tower. There were also no lights on the airplanes. At night we occasionally flew with battery-powered navigation lights, but we didn't go far away from the airport. At least I didn't.

AIR KNOCKER

I took all my training in an Aeronca Chief, before the Champ came out. When it came time for my solo, I was in a Taylorcraft. Not used to the glide ratio of the different wing, my first landing floated half way across the airport.

I would have been at Bishop Field then for about two years. After I gained permission to fly solo, without an instructor, the use of the plane cost seven dollars an hour. I made fourteen bucks pay for the week. Not much was left over for an instructor more often than was absolutely required. It was kind of a poor deal by the time I figured out my bus transportation and all.

But it was the only way to get your license. You never complained; you never went after a raise?

Oh, god, no. For this fourteen dollars, I propped planes, pumped gas, added oil, tied planes down on the ramp, put them back in the hangar, serviced them each night, changed tires, and did small fabric jobs.

The first fabric patch I put on, I had no idea what I was doing. The patch came out with a big lump. Ralph Rose, our instructor, teased me something fierce about it. "It might be kind of nice if somebody would show me how to do things around here." He got a big kick out of both my patch and my reply.

These two young women were always plotting and planning ways to include me in their fun and games. We used to do a lot of American Ace nonsense, chasing one another around in the sky.

We could do a few things a little differently than a pilot can now. There was nobody on the radio to tell us where to go or ask what we intended to do. We'd just get in whatever plane we could get our hands on and fly. Once in the air, if we wanted to fly from point A to point B, we would just aim the

Flint Breakfast Club, circ. 1945. Seated: Uncle Bulgie, 5th from left. Standing: second from left, Berniece Bowers; another student; Bob Vaillancourt; B.

airplane in that direction and go. We didn't have to mess around with asking Big Brother for a vector. As students, we often had three or four planes flying in loose formation.

I remember climbing up as high as I could manage one day in a little plane from the flightline. I didn't get to ten thousand feet, considerably less. If we got those planes near their maximum height, they were just about out of go.

I heard the talk about guys flying the hump in China. They told about the bad weather they were flying in, but I couldn't see where that would be any problem. So I took off one day in a Taylorcraft without any instrumentation. Going up into a cloud at a nice even rate, everything seemed easy and calm. But it didn't remain that way! I remember coming out the bottom of the clouds feeling upside down and backwards without knowing how in the world I had gotten there.

You sure lose your sense of feel when you don't have a visual cue.

The Taylorcraft had a cork bobber for a fuel gauge with a twelve gallon tank set right in front of the instrument panel. This cork was wired with a little hook on the end of it. When it got halfway down the marker, we'd head back to the field to get more fuel.

I always flew my two hours of pay when nobody else would fly. If somebody else was flying, I had to work. When the weather got real bad, I could get a plane. I'd fly in the forty-mile-an-hour wind, with ceilings five hundred feet above the ground (AGL), in rain and all sorts of nonsense. I wouldn't say I turned out to be a pretty good pilot from flying in those conditions, but let's say I got my time in anyway.

A private license only required about thirty-five hours of flying back then. The commercial took two hundred hours. A commercial license didn't entail anywhere near the things it does today, but we had to do spins then even for a private and they don't always do spins for a commercial license now.

That's right, Bob. We didn't have to put in any great amount of time to upgrade our license. Commercial only required cross-country for ten hours with three landings within a two hundred-mile range. That wasn't much training for the responsibility of carrying paying passengers.

The three of us used a very primitive algebraic equation to calculate the effect of wind direction. Berniece got all A's in algebra but she told Bob, "I just can't get this, I can't comprehend it, how will I ever get my private license?" So he showed her. She was so bored listening to his explanation. That was their first date. She was still just after him then only because I said she couldn't get him!

Berniece also used to go to the far side of the airport from where I was working and pretend her engine quit so I would have to go prop it for her.

The airplanes we were flying were not the best maintained in the world. They were all quite old and could have used a little honest repair. I didn't really understand the front end of a plane from the back end when we started dating. She could get me to do her bidding simply by pulling the throttle back to stall the engine. Something as simple as that can make those engines quit.

We had a lot of deadstick landings then, didn't we? People now panic if an engine quits. We didn't panic. Engine's gone? We landed, got out of the airplane, spun the prop ourselves and tried to crawl back in before the plane ran away.

We had conventional carburetors (carbs) then, which had the tendency to ice up with the slightest drop in temperature or with an increase in humidity.

*Berniece and Bob Vaillancourt: the Newlyweds.
Notice how tight the fabric is on this piece of their T'craft.*

The simple failure to pull the carb heat could cause it to ice up and easily stop the engine.

This was just one of a few poor things about flying back then. Perhaps a more fortunate thing for me was at the tail end of the war they had more pilots than they knew what to do with. While I was standing around waiting to go through their preflight course, they sent me through a maintenance program. That's where the rest of my life really got started. When the Air Force abandoned the Cadet program, they simply released me. I went back to Bishop and began working as a mechanic. Berniece and I married shortly after I got out of the service. We met when she was seventeen and we got married at twenty-one.

Air Frames and Power Plants

I worked again for a while back at Bishop, went on to Spartan in Tulsa, Oklahoma, to get an airframe and powerplant (A&P) license, and returned to Bishop. I've been in aircraft maintenance ever since.

After a while I went to work for General Motors Air Transport in main-

tenance and actually flew a bit for them as a co-pilot for a very short time. The pilot I replaced wanted to go somewhere else, but then he decided to come back, and I returned to spend the rest of my career with General Motors in aircraft maintenance, finally retiring as superintendent of aircraft maintenance.

Shortly after we got married, Berniece and I also started our own business. We had a couple of T'crafts. B used to lease them. Our planes didn't have any radios at first. B was trying to operate them by sticking her hand out the window to wave at somebody. We began to argue about how much money we were going to put into them in the line of radios, finally ending up with an Airboy. A couple of Airboy radios and then a Senior Airboy. The radios made us feel really uptown, very sophisticated. The tower in Flint was getting away from light-gun signals to control airport traffic. They demanded better equipment in every plane. But our engines were not even shielded to block noise for radio operation, so half the time the students couldn't hear what the tower was saying over the noise of the engine. Our radios were also strictly receivers. Pilots could not talk to the tower. If they had a problem the only thing they could do was wag their wings.

When I decided to open my own business in the fifties, a company only had to own one airplane outright to get a license to operate a fixed-base operation (FBO), but it was a giant step for me as a young woman to come up with even that much. In the hangar where Bob was working was a little Ercoupe that belonged to the general manager of Buick Motor Division. It had never been flown much by anyone but Bob and the General Motors pilots. It was going to be sold. I don't think it had ever been left outside a night in its life, so I bought it. Once I owned that Ercoupe, I could get the license to have the FBO, but I still needed airplanes to train students 'cause I sure wasn't going to let them fly my gorgeous, shiny Ercoupe. That's where Berniece and Bob came into my business picture. They bought two Taylorcrafts with the help of a fellow mechanic at GM.

The T'crafts we rented to B were in fairly good condition. At least they were better than the ones we learned in. Ours were nicely flyable; something to get some time built up in.

They were certainly flyable. As owner and mechanic, Bob had a vested interest in them.

As we worked in the Buick hangar, when the airplanes B used went taxiing past, our big joke was to say, "There go the little moneymakers now."

I couldn't even come anywhere near thinking about doing something to help someone like that now, with all the new liability costs. Just to tie the airplane down on the field would probably cost more money than could be made teaching students. When we owned airplanes to rent, there was no red tape. They call the new procedures the legitimizing of general aviation, but sometimes it seems they regulate and insure the industry to death.

Strains and Rewards

It was a tremendous financial strain on the Vaillancourts to help me get started in business, because they were young and trying to begin their family. As a young single woman, I never could have gotten a loan to start a business.

Our T'crafts weren't all that much strain. We were glad to help and earn a little extra on the side. But sometimes timing...if you can just get over a little hump, it doesn't take much more effort, if you are having fun. The investment worked out well for all of us.

To tell the honest truth financially, I don't think we even carried any insurance. If one of those planes ever got busted up, I would have been able to take care of the repair myself. By comparison, if we were renting in the crazy world of today and somebody busted a plane up, man-oh-man. The business would be eaten away just with the court costs. People can't do what we did anymore.

There are a dozen costs now that we didn't have then. The radio equipment today costs more than we paid for any of our airplanes. When B went into the Piper dealership, we sold all our Taylorcrafts. They had become obsolete.

Speaking of radios, Bob, remember that when we started flying, we rocked the wings to signal we understood the lightgun? If we continued to rock the wings, it indicated an emergency. We couldn't tell anybody what was wrong, but they would get emergency trucks going to meet us. When I lost a ski during a flight, I rocked my wings. Much later, when I did have two-way communication, I was flying a Twin Bonanza owned by Janey Hart when its cowl came loose. When I looked for the microphone, it was nowhere to be found. I rocked my wings. Fortunately, the staff in the tower still remembered this meant that I had an emergency! It felt strange to me then to be flying a large modern airplane while obeying the old light gun. Changes like this helped reinforce my interest in the place of people in history.

A Bridge of History

What a person is able to accomplish in one life is almost always a matter of timing. During World War II, when I was growing up in Flint, Michigan, it seemed that events overshadowed individual people. When I became serious about aviation near the end of the war, I became part of the bridge between what is called the golden age of aviation and the beginning of the space race.

To bind my story to its place in history, in 1925, the year I was born, Coolidge was president, China's Sun Yat-Sen died, *Mein Kampf* was published, and France began to build the Maginot Line for protection from German invasion. Synthetic rubber was developed. Bennington College for Women was established. People gathered to share fun more often in public places: in silent films Charlie Chaplin was capturing hearts and in Paris the Charleston was introduced. Demands were made for easier travel conditions: the first motel opened and Ford inaugurated commercial air service between Detroit and Chicago. One year later Stalin became dictator, Mussolini assumed total power and Emperor Hirohito began his reign.

When I graduated from high school in 1943, Franklin Roosevelt was president, the war turned against the Axis, Mussolini resigned, the United Nations was proposed, the Pentagon was completed, and the Jefferson Memorial was dedicated. Although the ban on non-essential driving was lifted in the United States, shoes, cheese, flour, fish and meat were still rationed. GI Joe began acting out the macho dreams of little boys. ABC began broadcasting. The popularity of *The Fountainhead* and *The Little Prince* indicated a new intellectual hunger. *For Whom the Bell Tolls* was made into a movie, and Jitterbugging replaced the Lindy Hop in popular dancing. Oral contraceptives also were introduced then, but it took about twenty years for them to gain acceptance. It took the same amount of time for Rachel Carlson to realize we would have a *Silent Spring* if we didn't ban the 1943 savior of crops, DDT. Nearer home the stress of Detroit race riots was somewhat balanced when the Red Wings won the Stanley Cup. But few of these things directly entered my everyday world.

Closer to home, the image of women who were Air Force Service Pilots in slacks on the silver screen fought with my image of Mother, on her way to grocery shop, wearing a hat and gloves. She loved to have neighbors over to our house, but seldom went to someone else's house. She gave up driving a car after she had a small accident. Then she began to wait until her husband, Ray, could take her out, anywhere. If she ever dreamed of selling her wonderful home-baked cookies, her loan application would have required her husband's signature.

Approach Pier: 1945

During the war, new two-place training planes were only produced as bird-dogs (spotters) to fly low over enemy positions for the war effort, or grasshoppers (couriers). These war years represented a real leap in the development of aviation equipment, but it took a while to have those changes available to general aviation. Ford built the first rate-of-climb indicator in 1942, but it was only for military use.

When I started to fly in 1944 and until after the war, we had access to Taylorcrafts produced up through 1942. Our T'crafts were equipped with an airspeed indicator. A bobber acted as the fuel gauge. One dial was both tachometer and indicated oil temperature. We also had a compass. We could tell how high our plane was above sea level. The altimeter we depended upon was a non-sensitive, one-needle instrument. The dial had marks like a clock and a single needle, but no minute hand. Using this instrument was like telling time with only the hours indicated. Thousands of feet were shown, but that wasn't much help in judging the hundreds of feet left before we landed, smoothly or with a plop, based on our own judgement.

Although our compass was a mandatory piece of equipment, nothing said it had to be dampened, or that it even had to work. A wrench in a pocket could swing the compass, a watch with a radium dial could also swing it. It would swing either side of our course in rough air. Until compasses were dampened, we were forced to learn how to fly at an angle across the section lines they used for surveying the state. These section lines became the basis of many of Michigan's roads, saving novice and experienced pilot alike from many an unwanted mile off course. One degree off course is one mile off course for every sixty miles flown, so it doesn't take long to get lost in an airplane. It wasn't until I flew outside the state that I learned roads might not follow exact north-south, east-west compass directions. This challenge tested my dead-reckoning skills anew.

Without specialized instruments anytime we left familiar territory we had to make exact calculations and we might have to revise our angle to the wind at every checkpoint. I remember one time very early in my education when I was flying with my instructor, Ralph Rose, on a dual cross-country. I had to prove I could fly solo cross-country and return without any assistance from good old "Uncle Bulgie". The plane was somewhere over a marshland with no points of reference when I realized I wasn't sure where we were. About the time I became so worried I thought I was going to be airsick, I noticed Saginaw Bay. Without realizing it, I already had the T'craft headed in the general direction of home. Fortunately, Ralph understood my moment of worry and I was released for solo cross-country without the expense of a second dual flight.

The use of a radio in the airplane was absolutely foreign to us. Eventually we had receivers, but it was hard to crank in the right frequency on the dial. These receivers were not shielded, so we could hear the engine running better than we could hear any communication. The low frequencies assigned to us also reported best any thunderstorm within miles of our location. When the early radios were on, they moved the compass off the magnetic heading we had selected. It is easy to understand why I never fully relied on radios. But this lack of instruments didn't stop me from flying long distances. Because I grew up in the early days of unreliable radios, I understood the risk/gain choice Amelia Earhart made when she got rid of the weight of her antenna on her round-the-world flight. I don't think it was a smart decision on her part, but I do understand it.

As students in Flint (FNT) we paid seven dollars for the plane and three dollars for the instructor. For comparison, at our local community college (NMC) in Traverse City (TVC), students now pay forty-two dollars for the most basic plane (Cessna 152s) and fourteen dollars for the instructor, who is likely to be someone trying to build time to enter some other form of aviation.

If I wanted to fly in Michigan from FNT to TVC after I earned my first license, I'd start my engine, watch for the green light from the tower, taxi to the end of the runway, watch for a green light again, fly via whatever route I wanted to take north to TVC. Once I was at TVC, I'd fly diagonally across their runway at a high altitude, read the wind direction from the tetrahedron and circle the field at pattern altitude until TVC Tower gave me a green light to land. If the weather was predicted to be poor anywhere along the route, and I couldn't fly around it, I'd simply stay in FNT or TVC until I was sure the bad weather had moved off my flight path. That kind of waiting is how Berniece and I wound up at the Wildcat Convention in Alpena without a secure room to sleep in.

In the period from the mid-40s to the mid-50s when I flew at Bishop Airport, the arrival of something as simple as the heated airplane was a neat thing. Sound deadening attempts and more reliable powerplants were also reasons for celebration. Eventually little wind generators operated navigation lights, which were laughably unreliable. After the war, when new instrumentation became available to us, it was designed for bomber-sized airplanes, all huge and heavy. Radios would be controlled in the cockpit, but the bulk of the equipment was located in the rear of the airplane, which added a new weight and balance factor. Still, we were happy with each step, until they developed something better. When Lear developed the first autopilot for general aviation planes, it was a real triumph. Taylorcraft started building aircraft for general aviation again in 1946. My A&P mechanic friends purchased these later T'craft BC-12Ds to rent to my business, Trimble Aviation.

Exit Pier: 1966

Flying cross country can be done without the sophistication of full instrumentation or the modern array of radio equipment, but in serious racing it was an integral part of the race equipment. In my last years of racing, my Piper Comanche panel had a clock, airspeed indicator and directional gyro. Those early directional gyros were great, but they would lose gyration and had to be reset constantly as we flew across the country. The Comanche also had an instrument that created an artificial horizon, and a Very high-frequency Omni Range (VOR) navigation indicator. Every time I have to say VOR, I'm glad for acronyms in aviation. Flying across country from coast to coast, I was also glad to have a manifold pressure gauge, tachometer, two-way radio and could tell if all of our flight instruments that depended on air-pressure were reliable (rate-of-climb indicator, turn-and-bank indicator, and altimeter). In addition, the Comanche had retractable landing gear controls and eight fuel, engine and oil gauges on the panel. All of these improvements let me fly faster, more sure of my destination.

The gauge that created the artificial horizon was especially important. When I raced, I made sure the little silhouette of my airplane, from wingtip to wingtip, was parallel to this artificial horizon. The minute an airplane comes "off the step" of straight and level flight, airspeed decreases. A pilot can't let an airplane do any flying by itself and expect to win a race. On the windshield above my panel I could glance at a calibrated compass (accurate, but still slow to respond to changes in heading). Below the panel ran a row of twenty different buttons, knobs, plugs, fuses and the radio controls. During a race my co-pilot and I paid special attention to the entire panel. If I raced in this new century, some of the most advanced equipment I used to win my last race would be considered archaic.

On My Own

I actually knew I wanted to fly long before that first ride in the Condor with Mr. Chamberlain when I was twelve. The dream began shortly after my father died in a house fire while trying in vain to save my brother and sisters. Mother had a rough time adjusting to this terrible loss. When it was especially difficult for her, I occasionally lived with relatives. This changing center of life started when I was one and a half. But even in her darkest times, mother did her best to help me. When I was a little older, she came home from a trip with a pedal car for me that celebrated the *Spirit of Saint Louis*, the plane Lindbergh flew across the Atlantic. That gift was the first in a series of events that began to direct my life onto a different course. As I grew older, I wanted more and more to make my lost family proud of me. I vowed never to accept mediocrity. In high school, I never imagined I could get a job flying, but I knew I wanted to become a pilot. While other little girls were playing with dolls, I was

It was the Spirit of Saint Louis to me!

building model airplanes. When my friends were going off to college or getting married, I continued my love affair with aviation.

Pilots in the news always seemed very exciting. They were not particularly physically glamorous, but as a child, I thought they were gods. All of the pilots at least wore clothes that were different. The early pilots were especially showy. Roscoe Turner set a US speed record for a distance of a hundred kilometers in Detroit by flying 289.908 miles an hour, maintained a handlebar mustache and traveled with a baby tiger. He won the Bendix Race in 1933 and won the Thompson Trophy more often than anyone else. Roscoe was a pilot to dream about.

In Flint, we wanted to stand out as pilots, too. We had to have big wristwatches. Living at home, riding my bike to the airport, I finally saved enough money to buy a Chronograph watch with all the circular slide rules. To me, that was *the* symbol of being a pilot. Then I had to have Ray-ban sunglasses and a padded bomber jacket of blue serge.

Eventually I could even afford the A-2 leather fighter jacket. We could buy the Real McCoy at the surplus store inexpensively after the war, so we all had them. I don't remember much insulation in them, but I can remember wearing them all winter. I also had the long white silk scarf because it was warm. These were practical things, but also gave a little notice of who we were...and I *liked* who I was!

Student or instructor had to be thin by the time we got our winter stuff on. To finish the list with my winter gear, I wore wool slacks with wool kneesocks, leather mittens with gloves inside them and big sheepskin boots. It's amazing to think back to those great big bomber pilot's boots searching the floor for the two-inch round brakes of a T'craft. We could just barely fit our heels on them. Even with all of that clothing the sixty-five horsepowered T'craft felt a little colder inside than it was outside. We could pull the heater knob without feeling much effect. Only in the summer could we tell the difference with it on.

Career Choice

I learned it was possible for women to make flying a career when I met a flight instructor, Margaret Crane, who freelanced at several small airports in southern Michigan. A friend of Margaret's, Fran Bera, was not only instructing, she owned a fixed-base operation (FBO) over in Battle Creek. Immediately after I met Margaret and Fran, I knew what I wanted to do with my life, if only I could fly well enough. They gave me the drive to go for a commercial license. To combine my love of flying with making a living seemed almost too good to be true.

Two hundred hours were required for a commercial license. An article in the Detroit paper indicated that Canada urgently needed flight-trained people, men or women, for their ferry service. This presented an exciting possibility so, with just over sixty hours of flight time, I dashed off for a visit to the Canadian Military Office in Windsor, Ontario. My hopes were high as I entered the very old, prestigious-looking offices of the enlistment branch of Canadian Services. With typical British tact, I was handed an application to fill out on the spot. After filling in the necessary information I noticed, in rather small type on the bottom of the form, that the term of service was "until the King no longer desires your services." That seemed a rather open-ended contract. What would constitute the undesirability to which it referred? Asking shed little light on the subject and, thinking about the state of world affairs, I decided if enlistment in the service was possible, it really should be for my own country.

In the early forties a group of women pilots was selected to ferry U.S.-built aircraft to European bases. Nancy Harkness Love led this group, called the Women's Auxiliary Ferry Service (WAFS). A wealthy socialite, she began to learn to fly at sixteen and started selling planes in the mid-thirties. Later, she married a competitor. (She must have been doing well if he wanted her attractive competition to be on his side of the fence.) As the president of Boston's Intercity Airline, Nancy's husband won a contract to fly planes to the Canadian border. Nancy made some of these deliveries. Adela Riek Scharr quotes her in *Sisters in the Sky, Volume I— The WAFS*: "I landed at an airport right at the boundary of the United States. I taxied the plane to the border, cut the switch, and climbed out. Then some Canadians pulled and shoved the airplane across the border. I turned around and brought back another one." After doing this for a while, Nancy wrote the Army Air Corps, stating she knew forty-nine women qualified to ferry planes for them. Interest in Nancy's idea slowly began to get off the ground when a shortage of male pilots began to develop. The basic idea was to relieve men for combat duty; a gallant offering, if only in appearance. Still, since women were normally forbidden to fly U.S. military aircraft, those women who met the qualifications jumped at the chance to fly the *big* ones.

A little later, another group of women was being screened for flying special assignments. Jackie Cochran formed this group, the Women's Airforce Service Pilots (WASP), when the Corps needed even more ferry pilots so men could be moved into combat. Jackie won the Bendix twice, earned the reluctant respect of pilots from around the world with her firsts in altitude and speed, was honored with the Harmon Trophy at least four times, took US women to Britain to fly for their war effort and rubbed shoulders with presidents. Her WASP was made up of initially less experienced women pilots who were equally as motivated as the WAFS pilots. When she developed the teaching-recruitment program that I still admire, Jackie applied techniques from the school of hard knocks that led to her own success. The WASPs went through basic training just like the men. Then, just like the men, they were given specs for a plane as they climbed into a new cockpit and, just like the men, they had to deliver it to the required destination. The result was a successful delivery record as good as the men, with no more casualties.

Many of the deliveries the WASPs flew were ones men did not want as long as combat service was available to them. Another of their gold-plated jobs was towing gunnery sleeves for target practice by new, would-be military pilots, who happened to all be men. The hazard value of these jobs sometimes equaled, or surpassed, that of the combat pilot. Still, it is not surprising that the women who answered this call of duty developed a feeling of being on equal terms with the men.

Perhaps this ego satisfaction intrigued me. Whatever the reason, I was certainly going to try to be one of the elite that would have the opportunity to fly the aircraft featured in every newsreel. I spent every dime and waking moment preparing to join those women as soon as the required hours came within my reach. But by the time I had the hours, the WASPs were discharged and I found I was competing for a job with male pilots home from the war.

Storm Clouds and Silver Linings

It was nearing the holiday season in 1945, Thanksgiving Day just a week away. My fresh new private license was ensconced in a brand new leather license-holder. My flight time was beginning to add up. Jean, the secretary for Francis Aviation, was up to her usual antics one day when my instructor, Ralph Rose, asked her to do some paperwork. She refused, starting to tell him just exactly where he could go. Apparently that was neither the place he wanted to go, nor was it the right time to refuse his request. I didn't stay around to hear the end of the discussion, as it was past time for me to head home. The long ride through Flint on the bike gave me the time to reflect on the day's happenings and to imagine how this little argument could put the whole flight school in a precarious position.

Late that evening, Ralph called to ask if I would like to take Jean's job. The argument had not ended with my departure. It ended with Jean's leaving. His phone call went something like this: "Do you type? Can you keep books? Are you willing to sit at the desk and schedule students and keep track of the money?"

Since I originally wanted to become a doctor, the courses I took in school did not prepare me for office work. I had to answer with the truth, but still get the job. I said I could learn to type and keep books. I became especially motivated to learn to type and keep the books when Ralph said I could take half my pay in flight time at a reduced rate. He also would let me take any special flights I was qualified to fly to build flight time toward my commercial license. By then, it sure didn't take much to convince me to stop smelling the oleum at A. C. Spark Plug. I also slowly began to realize that flying was becoming my life's passion.

Then I learned that Jerry Francis, for whom Ralph and I worked, was a hardheaded businessman. He required a full report every week. It took me all week to type it up. No one cared how long I took as long as it was done, on time, and accurate. Every spare minute I had was used to get flight time. For each hour I was secretary, I was paid something, plus two hours flying, plus I earned time in my logbook by ferrying planes to Lansing for maintenance. With my new status as office manager, I also earned a new name. One of the mechanics began to call me "Miss B". Soon it was picked up by everyone: my boss, the students, my friends and family. (To this day, my best friends know I prefer to be called simply by the letter "B".)

Soon enough my next big day arrived. The federal inspectors were coming to Flint to give the written examinations for the commercial license. I'd prepared for months, read all the books, attended ground school and felt ready, but apprehensive about taking the exam. After I spent the maximum amount of time allowed to write the test, I flew the airwork portion of the test from the airport in Romulus. Then the only thing I could do was impatiently await the result. Ten days later, commercial license in my not-so-new leather wallet, Ralph met me as I returned from a celebration flight and asked if I would be interested in working for him now as pilot-in-command, doing office work part time. This sounded like a great chance to continue flying. I also knew any job flying for hire was hard to find. So I quickly and gratefully nodded approval and Ralph introduced me to my first commercial passengers, an older couple who had missed their connecting flight to Florida. They needed to get to the airport in Detroit as soon as possible. We climbed into the assigned plane, an elegant four-place Luscomb sedan, and the next phase of my career began.

When I finally returned home, having passed my commercial, Ray was so excited he told me to invite all of my friends over to the house. My stepfather didn't drink. I don't think there was ever any alcohol in the house when I was growing up. At least I don't remember seeing any. But for this party, he bought cocktail glasses and booze so we all could have a celebration. It was a turning point. My piloting friends were always invited to drop by the house from then on.

Lengthing Hours

Out at Bishop, I didn't do very much charter work at Francis Aviation as long as they only had T'crafts. Those weren't even too great for hopping passengers. As the war wound down, we began to get war surplus Fairchilds, the open cockpit models. These were primary trainers for the military. Our flightline now included a plane with a one hundred forty-five horsepower radial engine that still took an inertial starter to get going. In the wintertime, having to crank this thing really took something out of a person.

We were always busier flying in the summer than in the winter, but we were also busy in the winter with all the extra effort it took to get the birds in the air. Francis got the PTs (primary trainers), moved on to BTs (basic trainers) and added ATs (advanced trainers) by the late Forties.

Now my job was to keep the planes flying by whatever means I could. I continued to work at Francis Aviation through getting my instructor's ticket and instrument rating. Ralph, who was still my good friend Uncle Bulgie, was gone before I got my instructor's license. When he left, Gertrude Prochaska took over.

On my way to CFI.

So Flint Aviation Takes New Glamour
Two Girls Teach Flying Students at Bishop

By Harold R. Gerace

Glamour has been added to flying lessons at Bishop Airport.

It's furnished by two school marms—flying school.

This pair has successfully invaded what at one time was a strictly a man's profession. They are Gertrude Prochazka and Bernice Trimble. Both are flight instructors for Francis School of Aviation.

Between them they've soloed nearly 100 of Flint's present crop of pilots. What's more, many of the masculine fliers prefer them to the men instructors.

Gertrude—Mrs. Prochazka (darnit) — is a half-pint with about 1,500 hours of flying time. About 1,000 of those hours have been spent instructing. Bernice is about 700 hours in the air. She isn't married.

Only 24, Gertrude lives at 426 Rasch Ct. Rasch is her maiden name and the street was named in honor of her dad who donated the property to the City. She is married to Laddie Prochazka who is flight engineer on Buick executive's twin-engine plane.

Yes, it was an airport romance. Gertrude met Laddie about two years ago when she had just got her instructor's license. Last October they were married.

She has soloed about 60 students and strangely enough, she prefers to teach girls how to fly. At present she isn't instructing. She does the office work but when Summer demands arrive, she'll be back teaching.

Bernice is only 22 and received her instructor's rating under Gertrude's guidance so she is sort of a protege. Living at 1609 Maplewood Ave., Bernice is a 1943 graduate of Central High School. She learned to fly in the Spring of 1944 and has been instructing for better than a year and has soloed upward of 30 fliers, all men.

These two girls, the only ones among about a dozen instructors at Bishop, both worked their way through flying school. After they obtained their wings, they did office work at Francis and instead of receiving pay, they got flying time.

From observation and talking to some of the pilots they graduated, it was learned they have a gentle but firm manner and their male students like it. At least they prefer it to the "cussing" out they sometimes get from the men instructors if they pull a bonehead stunt.

Both have flown the Army training ship named the AT-6, a 650-horsepower job, which is quite a setup from the usual 65-horsepower Taylorcrafts they normally pilot. Bernice has even flown the twin-engine UC-78 Cessna.

These two are intensely proud of their profession. They earn more money than the average job pays, for one thing. They receive a guaranteed minimum based on $3 to $4 an hour for instructing. If they do a lot of teaching, their pay is increased by every hour they fly above the minimum.

Gertrude says she doesn't intend to make a career of her work, but Bernice states simply: "This is my career."

Gertrude Prochazka (left) and Bernice Trimble, Flying School Teachers

Photo and text from Flint Journal (1947)

One day a most interesting man came into the office to get checked out in our Stinson AT-19, the military version of the civilian Stinson Reliant. Since this was a much larger ship than our training ships, I asked to see his license and logbook. He showed them to me, then began to tell of his most unusual background.

As a young boy growing up in Iowa, my new student always wanted to fly. He watched the men who came to Ottumwa to put on airshows with great interest. From the sidelines, he finally mustered up enough nerve to ask for a ride. Airshow people always needed cheap help with crowds, pushing the planes around and taking care of any other detail that was too dirty or too time consuming for the pilots. My new storyteller talked himself into the hearts of these showmen and was offered a job. Soon they were asking him to learn to wing walk. Knowing his parents' permission would be required, he made up a consent note. Without ever mentioning it to his folks, he started wing walking for the show. Everything was going along fine until the day the pilot lost a wing in the process of making a turn. The plane came crashing to the ground. The young wing walker was so badly hurt his life hung in the balance for weeks. Needless to say, his parents found out what he had been up to and, seeing his pain and the suffering he was going through without crying out, tried to accept this obsession that had captured their son.

By the time he allowed that he was the youngest wing walker in the country, the whole Francis School of Aviation crowd was entranced by Bob Millane. I decided if he wanted to fly the Stinson, I would surely try to make that possible. When I indicated I would be his instructor, he said, "I never had a woman instructor." We talked about performance and the handling characteristics of the plane, then walked out on the flight line, where he finally admitted he had very little experience with a Reliant. He did have plenty of overall flight time. He said he had flown a Stinson, but not the advanced military version.

Finally, Millane seemed to get over his reluctance to have me as his flight instructor, but I continued to be nervous, in part because the AT-19 was never intended to be a beginner's airplane. It had been used by the military as an advanced-navigation trainer. Francis Aviation bought it for the great interior space and its weight-carrying capacity. I loved to fly passengers and cargo in it. We could carry four passengers because the high wing had the ability to generate a great amount of lift. Powered by a three-hundred-horsepower Continental engine, it had a non-retractable landing gear, a controllable prop and a good range of operation. The Stinson only had one set of brakes, on the pilot's side. Unlike in the smaller planes, ground steering had to be done with a combination of rudder and brakes. The AT-19 also only had one set of throttle and prop controls, located on the left side of the pilot in the left seat. This meant reaching for them from the instructor's side (in the right seat) was next to impossible. So I didn't like to teach students in it. To top it off, I suddenly realized I knew less about this man than he knew about me. Although I'd seen his logbook, I really knew nothing about him or even if, in fact, the logbook and license were really his. Still, I decided I couldn't back down from the risk of Millane as a student even in the AT-19. In due time, he was able to fly the Stinson every chance he was given, because he was quite a competent pilot.

A marvelous character, Millane was also good-hearted, totally interested in flying and flying people. The accident early in his life damaged his legs and caused his upward growth to stop. He was five feet tall, a most robust man weighing some three hundred pounds, solid as a rock, with many talents. For example, he used a record player and special speakers, equipment that he put together himself, as a disk-jockey for school dances, wedding receptions, or any other occasion where someone was willing to pay him.

One day he suggested I create a booth out on the flight line near the hangar. Each Sunday he would set up his music and I would fly over town about the time people would be getting out of church, running the prop back and forth from high to low pitch, making a great deal of noise. I could see cars head for the airport. Probably they thought a plane was in trouble and hoped to arrive in time to see what would happen. Once there, Millane lured people out of their cars with his music. Then it was easy to convince them to buy tickets for rides. It was nothing to fly several hundred people over the city during one weekend, and it was well worth the work. Most of these people were making their first flight and their enthusiasm ran high. When we developed more passengers than I could handle, we added planes and pilots.

Hopping passengers was, at times, very exciting. One day a family came out to take a ride. The previous few Sundays the weather had not been good for passenger hops. It was getting late, so I was interested in closing up and going home, but the father had made a promise to his children. I hurried them along. It was getting dark and the wind was coming up, indicating the approach of weather. In a short time we were in the air, where I began telling them what the lights were on the ground. Suddenly from the back seat came a scream, "Mummy's dead, Mummy's dead!" I turned to look and, sure enough, the lady was slumped over. I said I'd head right for the airport and request an ambulance to meet us. The father told me not to worry. "She just fainted." I was to go ahead with the flight. He said she'd come around soon. I still took the shortest route back to the airport and contacted the tower. When I glanced over my shoulder again to calm the children, I noticed the lady was waking up. We landed. As they departed, I thought the man was a rather cavalier husband.

Another evening a group of people arrived and insisted I take them on a ride, right then. I suspected their impatience might be the effect of the alcohol they consumed to find the courage to take the flight. Their insistence and my reluctance drove me to ask an abnormally high price in the hopes this would discourage them. When it did not, I prepared the Stinson for flight, loaded my somewhat loaded passengers aboard and proceeded. I planned a short, routine flight. About the time the wheels left the ground, I felt a strong hand come down on my back, causing me to lurch forward on the control wheel. Before the wheels touched the ground again, I managed to level the plane and continue the take-off. Once safely in the air I turned to see what was going on behind me. Much to my concern, the biggest guy had unfastened his seatbelt. Our near accident came when he grabbed for the back of my seat, trying to see better. His whack nearly forced us into a nasty nose-down accident. I was no match for either of the two couples but, fortunately, they were now sure we would crash. The thought sobered them. With no further attempt to get out of their seats, the balance of the flight was uneventful.

These all were fleeting moments in my rather active lifestyle. I loved almost every long hour of it. Just as Millane predicted, even with our paying forty-three cents a gallon for fuel, the weekend passenger flights were a financial success. His showmanship was a big business plus, and our employers were quite satisfied. The camaraderie at the airport was increasing so we felt like a family, each interested in the others' welfare.

New Toys and Halloween

Then the old wing walker came flying in with a new toy, a BT-13. This basic military trainer seemed huge compared to our regular training planes, but sitting beside the Stinson advanced trainer, it appeared to be part of the family. Bob's BT-13 was a two place, tandem-cockpit aircraft with a four hundred and fifty horse-powered Pratt & Whitney engine. Its speed was much greater than any of our other training planes. We affectionately picked up this bird's wartime nickname, the Vultee Vibrator, and learned why the mechanics said it was thirty thousand nuts and bolts flying in loose formation. Even with a bit of teasing, Millane and his plane enjoyed the envy of our airport family for a long time.

One Halloween this family, i.e. Francis Aviation students, mechanics and spouses, decided to have a party in the hangar. Bob would furnish the music from the hundreds of records he'd collected over the years. A select group of students and their instructor, Miss B, decorated the hangar appropriately for the occasion. We gathered pumpkins and cornstalks by requisition from farmers we knew. It was nearly midnight when we finished. The hangar looked like a haunted house and very much like a party was about to begin. We closed the doors, tied the aircraft down on the ramp and went home to sleep. The next day was spent flying with students and, in between flights, getting cider, donuts and the rest of the fixin's for a real hoe-down, swingin' party. Finally after the intense day of flying, the swingin' was fun and felt just fine.

Aerobatic Snapshots

The occasional airshow in our area, as well as glimpses of big shows on the newsreels, motivated me to learn aerobatics. As a teenager, I would go to the local shows and watch in awe. There were two gals who used to be a regular part of the show in Flint. In my opinion they were the living end: Betty Skelton and Caro (Bailey) Bosca. Betty had a small biplane, a Pitts, named the "Little Stinker," and Caro had something similar, if not identical. With flawless grace, they each performed intricate maneuvers high in the air above the crowd. Of the two, I thought Caro was the better pilot, but the maneuvers of both pilots grew to embody freedom for me.

Once I earned my first license and built up some air time, Bob Chonoski, an advanced instructor for the Air Force, gave me my first taste of aerobatics. He demonstrated some maneuvers in a Fairchild PT23. Powered by a radial engine, that Fairchild sounded great! However, my first aerobatic ride was almost my last.

Bob flew us south of the airport practice area, climbed to about 3000 feet (today aerobatics are practiced at higher altitudes than this), then asked if my seat belt and shoulders harness were fastened. Well, they were, but they weren't very tight. He rolled us to inverted and I came out of that plane it seemed like three feet, grabbing everything I could get my hands on.

We had no radios, no intercom. Bob spoke through the gosport, "Let go of the airplane." I shook my head. (He was behind me.) "I'm not going to roll it back over until you let go."

Well, if the darn fool was going to kill me I might as well get it over with. When I finally let go, I realized why he had scared me. The seat belt will hold the pilot and, especially in aerobatics, the pilot has to rely on it.

It has to be cinched properly or, inverted, the rider will be out in the slipstream. I began to feel a real desire to learn aerobatics when I realized how secure the harness system was.

Soon I decided I was going to learn enough to enter the airshow business. I wrote Sammy Mason, a nationally known aerobatic pilot, to ask what maneuvers I should do. Dear Sammy, who didn't know me from anything, took the time to write a long letter about aerobatics, the airshow business and how to get started. He was one of the first in a long line who encouraged me as a professional.

Part of the practice procedure was to select a clearing with a wooded area beside it so the woods became the make-believe grandstand. I was to do all my maneuvers in front of this crowd. Sammy warned me to never start maneuvers headed down, but only with the nose of the plane pointed up into the sky. I practiced some fancy flying Sammy laid out in the letter and, boy, I thought I was getting pretty good. I felt I was no longer a danger as the plane under my command looped, barreled, snapped and slow-rolled its way across the sky.

Time to show off to my best friend, Berniece!

Cub and Me

My best friend, the PT-23 and the new aerobatic ace took off into the wild blue yonder. The ace did just fine for a while. Then, in the middle of one of my intricate maneuvers, we blew a rockerbox gasket. The engine started to throw oil over everything and anything in its path, including both windshields. I was in the rear cockpit. Since I couldn't see through my windshield, I undid my seatbelt, got up into the slipstream as far as I could and flew straight back to the airport. It didn't matter whether the pitot tube was plugged with oil; I couldn't read my airspeed indicator anyway. This was still before radios, so I couldn't tell the tower about my problem. I rocked my wings. They gave me a red light, so we had to go around. The next time they gave me a red and green light, which meant "proceed with caution." So, we proceeded to land as well as we could with me hanging out the left side of the front cockpit.

The minute the wheels touched the runway I could hear Berniece screaming. Suddenly I saw a J3 Cub right in front of us...in the middle of the runway! There was no way for me to miss it, especially in the wintertime with an instructor standing on the snowbank to our left.

The Pilot!

Here was the Cub, there was the instructor, and, on skis, I couldn't stop. I had to make an instantaneous choice because I couldn't miss both obstacles, so under the wing of the Cub I went. The pitot tube on the leading edge of the Fairchild hit the strut of the Cub with such impact it was bent straight up, cutting through the wing of the Cub all the way to the spar. The Cub raced along with us on down the runway. There we were, just screaming down the runway. I looked right into the face of the guy in the Cub. His eyes got bigger, and bigger, and bigger. Stuck to two women. He couldn't believe it. I saved the life of the instructor but nearly caused his student to die of a heart attack.

Moving On

I eventually met Dale Hath, the manager of Flint Aeronautical. Their flight service competed at Bishop Airport with Francis Aviation, where my friends were, but Flint Aeronautical was a Cessna dealer. Cessna was beginning to build bigger, better airplanes all the time, so when Dale invited me to, I started flying for them and got to do some demonstration work as well.

I also did a lot of ground school instructing. After Flint Aeronautical bought an old Link C-3 trainer, I went for a Link instructor rating. All simulators were called Links then because Link was the only one making them. These simulators worked on air bellows, which were actuated by an electrical system that was pretty complicated. The flight examination for

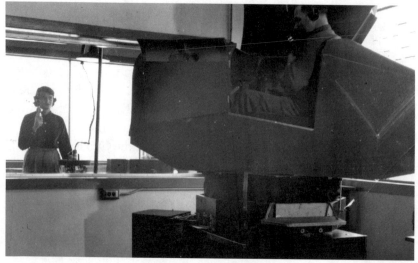

Miss B with Link trainer.

an instructor involved the mechanics of what made the instruments work in the unit, as well as the appropriate understanding of how the tracking crab recorded the student's flight path in the simulator. As much as I still admired my old instructor, Ralph Rose, I found I could not be quite as clever by using up bad weather playing red dog poker with my students. With the Link, I could have a student make an appointment to come regardless of the weather. Flying a simulator became part of a much better learning experience for my students.

When students are out in the airplane they are so conscious of attitude changes, they divide their attention between the airplane and the outside. A simulator allows them to concentrate. Also, the simulator was a much friendlier way to learn some things because we could stop students in the middle of a flight, show them their mistake and talk them through the correction. I feel a simulator is something every flight school ought to have because it is such a satisfactory teaching tool. Of course, it wasn't all serious. Once in a while an advanced student would be doing so well we'd light up a cigar, put it at the front of the intake and say, "I think you're on fire." The hood would suddenly be thrown open with the student boiling out. Or we would say we were going to do spins and set a student up for vertigo. Vertigo is a condition of disorientation caused by the fluids of the inner ear. Some people can handle recovery easily and some are disoriented for a longer period of time. When you are the student inside, all this simulated stuff becomes real. You can feel the spins: suddenly when you can't tell if left is right or right is left, you can get airsick. The instructor can do things that visually cue the senses, and it's dark, so the imagination can take over. When we realized a student was ready to begin to take in what was going on, we made sure something was going on!

Those cigars remind me of Turk Gillespie. He gave me some twin engine flight time in a Bamboo Bomber, always with an unlit cigar butt clenched between his teeth. He transferred to Flint through a program of civilian instructors for primary military training. The Bamboo Bomber was more officially known as the UC78, or a Cessna Bobcat. Turk flew this plane for Michigan Central Airlines. One time in the fall we were flying the Bobcat in slush and ice, landing from every angle and attitude. "Don't land in the water... splashythump!... your brakes are going to ice up!" Another time he turned the gas off to one engine. That made me jump on the rudder so hard my heart didn't have the chance to speed up until I got the gas flowing back into that engine. The torque, or tendency to pull off center with an engine out, was hard enough to correct that another nickname for that twin was the "Useless 78." It was a barely controlled glider. I felt cutting off the gas was a dumb thing to do. Turk thought it was funny. We laughed together a lot after I got over the shock of each challenge.

After the War

When the guys started coming back from the war, military pilots were given priority in being hired and they would take anything. These men really did not want to flight instruct, but they wanted to add up hours so they could compete for other jobs. Most of them were doing a miserable job as instructors. It kept a lot of women out of flying who wanted to instruct. I figured if I wanted a leg up in terms of getting a job in preference to these men, I had to have something they didn't have to offer.

I took a month off during the winter to brush up on all aspects of flying as taught in ground school: meteorology, navigation, airframe, engine and aerodynamics. While holed up in my folks' cabin, I hired a young girl from the family next door to read all the technical books aloud to me. As she read, I focused my mind to visualize every detail in every system. There weren't many instructors who had all the ground and simulator ratings, so I got every rating there was. I wanted to keep my job.

As it turned out, I actually had a lot of the military people pay to be flight-checked by me. They had learned to fly in such a different way, with an emphasis on heavy planes, so becoming instructors in light airplanes took a lot of redoing. Some of the boys coming back from the war had never landed an airplane. Their hours added up, but they were parachutists jumping from the planes. It was a bit disconcerting to hear, "first time I've landed," when their log books read two hundred or more hours in the air.

I soon discovered I really didn't have to worry about many returning pilots as competition. They thought they knew everything, but I found they didn't really know anything about civilian flying. One day I told a fellow I was going to pull his boots off if he didn't stop overusing the rudder. He couldn't feel the rudder pedals through the soles of his combat boots. Finally, he took his boots off until he learned to finesse the rudder.

Somehow in all the steady effort, instead of feeling I was competing against men, I felt I was competing *with* them.

I found I actually enjoyed working with men most of the time, but there were times...! The inertial starters on all the Fairchilds had to have their flywheel built up to a certain number of revolutions per minute, then you removed the crank, climbed inside and hoped the plane would start. Once, when I got back from an appendix operation, I asked Bob Chonoski to help me get a Fairchild started. He responded, "No, if you're going to do a man's work, then you're going to do a man's work!" I did a man's work.

Things really got going great guns at Flint Aeronautical when Cessna began to use the Whitman landing gear made of spring steel. I'd just taken on a new student who said he was never going to solo because he couldn't land the airplane. "Well, yes, you can land the airplane." No, he was sure he was going to crash. "I'm going to show you just what this airplane will take." I bounced that plane all around the airport, demonstrating the roughest landings I could imagine. "Well, if it can take that, fine, but I did a landing like that the other day and I clipped the prop." The possibility of this was news to me! He finally did solo.

The Test Pilot

A new desire of mine was to be a test pilot. I still was attracted to the engineering end of flying. There were few women test pilots then. Teddy Kenyon was testing for Grumman, but she was testing instruments her husband designed so her opportunity was unique.

I began to test ground-based equipment for AC Sparkplug. (Didn't I say I would be one of the engineers for my old bosses in a few days?) I had to fly an exact two thousand feet AGL (Above Ground Level), with an exact heading straight down Center Road, pointed at the AC building. A camera mounted on a gunsight secured to the top of the building would photograph my flight. I flew that five straight miles over and over while they aimed the gunsight at me. Some days I flew in turbulence, some days in freezing ice. It took us close to a year to finish the tests, debriefing constantly while I was in the air. AC had to have a specific amount of footage showing the accuracy of their sights before they would be accepted by the military.

Another testing opportunity came when I began working for Flint

Aeronautical. Dale said, "I have a job for you. You're going off to New York. Take this airplane and see what can be done to get them to fix the tailwheel." I flew east to the factory in Lancaster where apparently the only complaint they had ever had was from Dale. When I flew in, they gave me a look that said, "We're going to put you through the mill to prove to you there is no problem." Instead, I demonstrated to them that landing in a crosswind in an airplane with their Scott pneumatic tail wheel created a shimmy that wouldn't stop. We could not dampen it. They had to figure out how to properly dampen this shimmy because it was causing ground loops. A ground loop can happen when an airplane seems to "spin on a dime" around the tip of a wing dragging on the ground. Students in ground loops destroyed airplanes. In some unfortunate situations students in ground loops can also destroy themselves. A week of testing, as they took the shimmy dampener apart to rearrange the springs, helped them develop solutions.

To teach students cause and effect, we did things with the training airplanes that a pilot normally wouldn't do, so we were often testing the limits of our aircraft. I don't think I realized the things I did then to solve the everyday problems my students had. It was all just a matter of getting on with business.

Loss and Recovery

One thing leads to another in life. I worked for Dale for quite a few years. As a matter of fact, we were thinking very seriously of getting married. However, one day when Dale was up with a student, another student collided with their plane through a terrible set of circumstances. There were no survivors.

The shock of that tragedy and his loss made me think twice about whether I wanted to stay in aviation. For the first time I was not sure of my goal. A friend worked for one of the downtown department stores. She wanted me to get into the millinery department where she worked. "Oh, just come down, and you can take over... ." Well, that sounded great, but when I got down there..."Oh, no. This is not for me!"

I went back to Flint Aeronautical as a manager for Al Koerts. We couldn't stand each other, so I again thought I'd get out of aviation. This time the magic of women in aviation took over.

Years before, I'd sold Ann Fruehauf one of the four-place Cessnas, a C182, when she and her brother were students of mine. They let me know they were flying to Florida. They thought I ought to go down with them to get away. So, I did. We had a ball. We ran into another former student and employee of mine, Bob Bowman. I introduced Bob to Ann, and they soon got married. Bob wanted me to stay and work at Embry-Riddle, the international flight school. I still didn't know if I really wanted to go back into aviation, so I returned to Flint. Soon after my return, Charlie Glidden ran into Dad down at the plant. He said he was looking for me to work for him. When I learned he only had one Aeronca and a little office, I felt I could try to work for him. However, I would not work for my former pay. He said to set my own price. In about a year of flying and managing students, I had Charlie's plane paid for and, with him as my cushion, I was happy to be back in the air.

Ninety-Nine Snapshots

Soon after I received my private pilot license in 1945, I was sponsored into the International Organization of Women Pilots known as the Ninety-Nines. Amelia Hendrickson (now Leslie), a local doctor's wife, invited me to join the Michigan Chapter at a meeting in Flint. The purpose of this organization "is to promote a close relationship among women pilots and to unite them in any movement that may be for their benefit or for that of aviation in general." In my case, they certainly have fulfilled their function.

Almost every book I read as a child on aviation and women was by Amelia Earhart, the first president of the Ninety-Nines. Early on I read *20 hrs, 40 mins*, the first book Amelia wrote. I've just finished reading *Last Flight*, published after she disappeared over the Pacific.

At eighteen, I felt in awe of several Michigan ladies who were pilots and members of the organization Amelia helped found. At that first meeting, my sponsor was the only one I knew, but it was clear there were more women in the group who I wanted the chance to get to know.

The Ninety-Nines organization has roots steeped in aviation history. Founded in 1929 by a small group of women pilots who were aware of the importance of forming an organization to gain a proper recognition of their skills, these women also had the desire to participate in the new breakthrough in aviation, transportation by air. A letter was sent to each licensed woman pilot in the United States, with an invitation to a meeting on Long Island. Ninety nine of them attended and/or replied.

The women pilots I know are talented and dedicated. As the years began to fly by, my Ninety-Nine membership became a golden key to my future. Sharing the experience of flight with other like-minded pilots was and is something I cannot value enough.

In one way or another, I'm sure these women could fit in as friends of yours: Mary E. Clark remains a very special lady in my heart. As you will learn in later racing chapters, she was the kind of friend everyone ought to have. Part of a large family, she was a devout Catholic. Her family owned the Crowley Boiler Works in Jackson, Michigan. When WWII broke out, Mary joined the Red Cross and spent the war years overseas. We shared many flying adventures.

Jean P. Reynolds was another Michigan Ninety-nine who became a best friend. She learned to fly in Jackson, becoming a flight instructor. At first we only saw each other at the Ninety-Nines' Christmas party. Later we took students on cross-country flights just to meet and have coffee and gab.

Amelia Leslie still comes to have lunch with me every summer. She is a lovely lady, a pioneering Michigan pilot originally from the Upper Peninsula. She lives in Florida most of the year now, so I don't otherwise see her. Our continued friendship is typical of the Ninety-Nines...we are there for each other any time, any day, and the value of the camaraderie is priceless.

Before I became the Michigan Chapter's chairman, I was chapter librarian. I set up a series of books to help women achieve higher ratings. Our library boxes also held inspirational books about women in aviation. I felt if we saw how they could achieve their goals, we could achieve ours.

FRIENDS TOGETHER: BERNIECE AND MISS B
(with Berniece Bowers Vaillancourt)

(Berniece) We both joined the Ninety-Nines right after we got our private licenses. One had to be eighteen to get a license, so B joined in 1945, first as usual. When I went with B to my first meeting, the Michigan Chapter wanted someone to take over as librarian. Being librarian meant taking this big stack of books to every meeting, so everyone would have access to them. The last thing in the world I wanted to be was the librarian! The first thing I knew, B was raising her hand, "Oh, I would like to be librarian!" So I had to help her carry all those stacks of books every month. We also had to rent the Stinson to have enough room for all the books. That cost more money!

That made both of us have a greater commitment to the Ninety-Nines. I added books to use for ground school study to this library because they were so expensive to buy. They became a popular item, but did add weight to the stacks we carried.

The second Sunday of every month, there was a chapter meeting somewhere in Michigan for all members in the state. I think there were fifty or sixty members when I was librarian. I remember our Christmas parties were at Alice Hammond's and were well attended. Berniece and I became excited about how vital all those women were, not just in flying, but in all areas of life.

One time we were invited to the Book-Cadillac Hotel in Detroit for a formal presentation to recognize important women pilots of the day. The Michigan Aeroclub was having a big reception for them. The Aeroclub was one of the oldest flying clubs in the country. We could only afford to attend because we lived near enough. Since we couldn't afford a hotel room, we drove down and stopped to change at the Y.

Previously I had never heard my friend swear, but suddenly, "Damn it!" "What is the matter?" "I brought one brown shoe and one black shoe." I just went into hysterics. But I quit laughing when B said, "I'm not going. I am going home. I am not going to go. We don't have time to shop for shoes." "Who's going to see you? You have a long dress on. Who's going to see your shoes?" Finally, she agreed she would go.

We had such a nice time. When I met Jackie Cochran and Blanche Noyes in the receiving line, I enthused, "We are so happy, so proud, to be here and to meet you. I want you to know how badly we wanted to come. My friend forgot to bring the right shoes. She is wearing one brown and one black, and we came anyway."

What a friend! All I wanted was the earth to open up and swallow me!

B was speechless, but those women loved it! We certainly stood out from the crowd! Blanche Noyes took us on a nice tour to Canada after the meeting. I am glad I never dropped my Ninety-Nines membership completely despite pressure from Mamma and the priest because I had two children. After I got back into the air, the fun continued, and the Ninety-Nines have been a private satisfaction all of my life.

RACING: MY FIRST TRANSCONTINENTAL

In some ways, I can credit Gertrude Prochazka for getting me into racing. Gerty in Flint, Fran Bera in Battlecreek, and Margaret Crane over in

Fenton were the only women I knew who were instructing in Michigan. I decided I was going to join them, so Gerty went to work. In the early to mid-forties, there weren't very many women pilots, so there was a question of whether anyone would want to take flight instruction from a woman. I knew how good we were compared to the men who flew, so I was surprised by this reluctance. Fran told us racing helped a great deal to become known as a professional pilot. By racing, we established ourselves as competent in the public eye.

And Fran was right; racing became good publicity for me. When I expected more competition from male instructors, I actually had much less. I never worried about gaining a good reputation as an instructor; it just sort of came. People from Lansing and Detroit were driving to Flint to take lessons from me. I had about as many students as I could handle. And I enjoyed it. I really did enjoy it.

When I first started long-distance racing we only had a directional gyro and depended on dead reckoning for navigation. That meant we had to understand the weather and our charts and interpret them correctly as we flew over the actual terrain. My first co-pilot, for the 1954 All Woman Transcontinental Air Race (AWTAR), was Lois Wilson, who is now deceased. She lived in the Dearborn area. I picked Lois, not that I knew her very well, but I had been watching her gain ratings rather rapidly, so she appeared to enjoy flying. I didn't know any women in the Flint area who were working as hard as she was. I thought it would be nice to have Lois because, as a pilot, she would be current. Well, she was nice and she was current on the ratings, but she was not current in having actual experience beyond the section lines of Michigan. And that was our only problem in the race.

Going out to the start of the race in California we had some rather jolly times learning to fly through the mountains, but it was on the way back, racing, that we faced the challenge of a desert storm. Have you ever encountered a desert storm? In the Mid-West, a storm cell will move from west to east pretty much at a uniform rate. Out in the desert they sit there and *expand*. It looks almost like a chain reaction. The cell in front of us was huge, blacker than black. When we finally got around it I turned to Lois. "Do you know where we are?" A blank, nervous face met my query. "Nope!"

I didn't know either for sure so I looked ahead and saw two little peaks. "On the way out we saw those in the area of Phoenix, but we're not going to Phoenix today; we're going to Winslow." We lost several minutes getting back on course, but taking the circular route kept us out of that horrible storm. This is the only time I was ever not sure of my position during a race. We didn't win my first transcontinental race, but we didn't do terribly badly either. I think we still placed in the top ten. Probably ninth, but we got there!

Facing reality, my co-pilot and I were both neophytes in terms of flying sophistication. Radios were still not too common. They only had VOR's (Very high-frequency Omni Range radios) to aid navigation. VOR's didn't work well when we flew low or when the station range was limited.

I never blamed my first co-pilot for the results of that race. Lois earned her ratings just like they treat students in some school systems now. If somebody is moved forward from the fifth grade into the sixth grade whether they understand what they should know or not, they are going to be lost at least for a while. My first co-pilot gained the ratings, but she didn't really have the knowledge she needed, nor the experience. But then, I guess as to experience, we were about equal. It was the blind leading the blind.

B. Steadman, International President of the Ninety-Nines, Inc. (Who me? The little girl with butterflies in her stomach as she rode her bike over the hill to the airport?)

Involvement in the politics of the Ninety-Nines really wasn't something that I coveted. I enjoyed the camaraderie of the group, but.... I was a Chapter chairman for a while, on the board of the All Woman Transcontinental Air Race, known then as the Powder Puff, and I raced as a Ninety-Nine. I did not come up through the ranks of the International board. I jumped in ahead a bit. I was not a secretary and I never thought of myself as being a treasurer. I went in as a vice-president, an elected position, when Dottie Young was president.

At this time Page Schamburger became one of my best friends in the Ninety-Nines. In preparation for possibly becoming president, I was supposed to come up with a theme. Page helped me refer to every meeting as a "fun and friendship" meeting, but we were noticing a lot of the history of the women in aviation was being destroyed. Ladies would die, and their children knew nothing about the importance of their role in aviation. Some of what was not being destroyed was...well, let's say the mind sort of has a way of making things look better than they actually were. Dates certainly got a little off. While I was vice-president, Page and I decided it was time to correct all those things, primarily to start people thinking in terms of preserving the real record.

Two years later, when I was sworn in as president of Ninety-Nines, Inc., the pilot of my very first flight, Clarence Chamberlain, sat next to me at the banquet in New York City at the Waldorf-Astoria. There was a year or two between the times I met him...many a year. That banquet was a grand event, but Clarence seemed rather long in the tooth in 1969.

After I became president, Page and I focused the membership even more on preservation. That focus turned out to be of lasting importance. In the beginning of our work together, whether we were going to have a place to put things was not as important as getting people to save their memorabilia, but we soon realized we wanted to get the record of women pilots' impact on history preserved for our own reference, and preservation of this record became a lifelong theme for Page and me.

Many people don't think that what they have done is important. They are simply too busy keeping up with life. The true picture of an individual isn't just what she or he does; it's what else is being done at the same time that makes it significant or not. Also, what one person has actually done might not be as important as her or his part in the total of what everybody has done to make something happen. Progress is built partially on the confidence of known history.

An example of the material we were collecting involved two people who were important in my own past. When I discovered Phoebe Omlie was living nearby in Ohio, Mary Clark and I went down to visit the old aviation pioneer and convinced her to come to Ann Arbor for the weekend, where I arranged for a wire tape recorder and Bob Millane to be on hand. Well! It wasn't until they got talking that I realized just how much they had in common. Early in her career, Phoebe was a parachutist, wing-walker and stunt flier. Bob, as an old wing-walker, drew out some marvelous yarns. While Bob was working with me at Flint Aeronautical, Phoebe was performing demonstrations in *Miss Moline* for Monocoupe. We shared a lovely, peaceful weekend.

THE SPIRIT OF SAINT LOUIS

When I was president of the Ninety-Nines, I was also active in Zonta International. Zonta is an organization of women (now people) who hold management or ownership positions in business or the professions. As another group of women who honor the mentoring work of Amelia Earhart, many are also Ninety-nines.

One time I was in Lansing for a joint Zonta meeting. Vern Jobst was the pilot of the *Spirit of Saint Louis* replica built to fly for the Lindbergh movie. The *Spirit* was promoting the film in Lansing along with a straight-wing Detroiter and, ... how in the heck did it happen?

While I was looking at the two planes, somebody came out and said they'd like me to meet Vern Jobst. "I've been wanting to meet you. I want you to fly the airplane." My mind began to reel, (Who me? Jobst has to be the greatest guy in the world to fly this bird!) This was just after I'd had some surgery, so I hadn't been flying in a long time. I hadn't even wanted to let on that I'd like to go for a ride in a plane, but I was not slow in accepting this invitation, assuming that I was going to ride along as a passenger. (This is really nice.) I got into the *Spirit* up front where Jobst indicated. (This is great.) After he started the engine, he asked me to taxi the *Spirit* to the end of the runway. (Now he's surely going to take off?) Then, Jobst yelled over the noise of the engine, "Take it off!" (Okay!)

Instead of having a normal windshield in front of the pilot, the *Spirit* was built with a solid metal front, the window space taken up with part of the extra fuel tanks. With no glass in any windows, the plane I flew was a true replica. The pilot had to fly the replica using only the few instruments Lindbergh used to navigate the ocean, or could look out an opening in the side of the plane, just like he did. Natural light from small unglassed openings on each side of the cockpit only made everything seem closer to me while we flew. How did the man keep warm?

As we started down the runway the *Spirit* felt strange and I felt strange, but all of it also felt really good. Controlling Lindbergh's plane on the ground was similar to controlling a Fairchild that has a radial engine. I couldn't see much to guide our direction, the rudders seemed too close and the control stick seemed too far away.

After I got the *Spirit* off the ground, Jobst urged me to play around for a while. So I did turns and a few stalls to get the feel of the bird. And it *was* a strange bird. They say the only reason for rudders is to take care of the unequal drag of the ailerons. Well, this airplane proved that saying was true. We'd only been up for twenty to twenty-five minutes when Jobst indicated that I had enough time in the air. Again I waited for him to take over the controls for landing...but he didn't. (Oh, man, I can just see the headlines now.) And as I was thinking that, I glanced over and saw Anne Morrow Lindbergh's signature on the inside of the fabric. (Oh, God, I can't crash this thing.) So I landed it! Then I taxied the *Spirit of Saint Louis* right up to a great big crowd of people which felt, well, just like the pictures of Lindbergh landing at Le Bourget.

There are two logbooks in the *Spirit*. One is the regular airplane logbook. The other is Anne Lindbergh's logbook of people who have flown the replica. So if you ever see the replica, ask to see that logbook and you'll see my little blurb..." Thanks for a trip back into nostalgia." I was so concerned with getting the historic airplane up and down safely I didn't even think much of what it would be like to fly the *Spirit* across the Atlantic. Only for a moment as we flew straight and level could I imagine the waves below my wings and think of the distance Lindbergh flew with no glass in the windscreen. Vern Jobst was a *very* nice man.

Carol pours the bubbly!

The Zontians in Lansing who were also Ninety-Nines were unhappy with me as the one who got to fly such a treasure, but Carol Leonard bought champagne to celebrate my flight and the whole group joined in!

As a child, I had a book written by Lindbergh with waves reaching out of a very black sea for his airplane. That image really impressed me. When he flew the Atlantic they had no weather observations except by ships at sea. Sea level and flight level can be in entirely different air masses. Lindbergh's replica seemed structurally sound, and I'm sure the engine I flew was updated, but as pilot that day over dry land, I was very busy.

I think pushing boundaries takes courage. I'm still amazed by his. People frequently fly small planes across the Atlantic and, you know, the ocean isn't any drier or any more shallow now than it was when Lindbergh flew across. Pilots are lost every year. Some of the airplanes people are flying across the Atlantic now don't look any better than *The Spirit of Saint Louis*. When Lindbergh flew as first, it took an extra lot of courage and determination. But if people really want to do something, they can do their best to prepare for it. Properly prepared, you're ready to take the risk. You know what the risks are, and you want to take the challenge badly enough to pay even the ultimate price. However, I don't think I would ever want to fly the entire Atlantic in a small airplane.

When the time came for me to prepare to push the boundaries, I learned that I did understand Lindbergh's willingness to take a risk. I also learned I was ready to do something that might seem even more adventurous than his flight across the ocean.

Racing: To Cuba With Joan

(with Joan Hrubec)

Having tested my wings by flying a few transcontinental air races, I decided to change from a Powder Puff Pilot into an Angel by racing internationally. (Believe it or not, that is what we racers were honored to be called back then.) This turned out to be a wonderful challenge.

Joan Hrubec and I entered the 1955 All Women's International Air Race (AWIAR) to Cuba with a Cessna 180. Joan was another child who realized model airplanes couldn't tell a boy from a girl. After adopting this attitude, she was the first winner of the Arlene Davis Trophy at a contest held during the National Model Airplane Exhibit.

(Joan) Arlene Davis was a racer, the first private pilot, man or woman, ever to formally become instrument rated. She was also the first woman to receive the 4-M, indicating she was able to fly multiengine airplanes with a gross weight over ten thousand pounds, above the land or the sea. The trophy named to honor Arlene was granted to me as the girl scoring the most points at the annual model airplane expo in Toledo, Ohio. Later, in 1959, Arlene flew the first private airplane to cross both the North and South Atlantic in one trip.

I met Joan shortly after she joined the Ninety-Nines, at the 1952 Michigan/Ohio joint meeting at Cedar Point. She drove a gray MG with a red interior! That made us soul-mates for sure.

Licensed in 1949, while a student at Stevens College in Columbia, Missouri, Joan owned a Tri-Pacer and had flown a number of planes from the Cessna series as well as an open cockpit biplane, the PT-17. Thus, I was fairly confident that Joan had a depth of experience that was missing in my first racing co-pilot.

Nineteen Fifty-Five AWIAR

When we flew to the start of the race, in Washington, D.C., I was to report our position to Washington National Control as the plane flew over the Masonic Building. I didn't have the slightest idea how to locate any one building in the maze that spread out below us. Finally we saw something big sticking out of the haze like a fang and reported. We were relieved when Washington Tower responded that they had us on radar where we said we were.

Landing at Washington National Airport really woke me up! Back home at Bishop Airport, when a DC3 took off from Detroit for Flint, we had to keep out of the way until the tower alerted us, using the old red and green light gun system. Washington National, second busiest airport in the world, required much more sophistication from the pilots and their airplanes.

Joan Hrubec, Miss B, and the Cessna 180.

The AWIAR, also called the Angel Derby, was developed by the Florida Chapter of the Ninety-Nines, with Blanche Noyes closely involved. By this time Blanche was the head of the Federal Aviation Authority (FAA) Airmarking Program. As one of the judges, Blanche pointed right at me as she said, "I don't want you to fly at more than seventy-five percent power." Maybe she remembered my brown shoe/black shoe incident from years ago in Michigan.

I was never in a position to have an especially groomed airplane for a race. Whatever I used was right off the active flight line. Later that night,

Blanche said she had been out to the airport, looking over the race planes. "I don't know how in the world you think you can win a race with an airplane with that many antennas on it." She helped us feel pretty special in our C-180, loaded with radios.

INTERNATIONAL AIR RACE
June 9th - 11th 1955

SCHEDULE FOR WASHINGTON, D.C.
Deadline Monday June 6th 1955, Hyble Valley

MONDAY, June 6th 1955
Handicapping - (Will be notified when your
Plane must be handicapped)
9:00PM, Pilots' Briefing, Willard Hotel

* * * * * * * * *

TUESDAY, June 7th 1955
Handicapping
7:00PM, Reception and Banquet, Willard Hotel

* * * * * * * * *

WEDNESDAY, June 8th 1955
5:00PM, Reception at Pan American Union
and
(early to bed for an early rise Thurs AM)

* * * * * * * * *

THURSDAY, June 9th 1955

4:30 AM	Busses will pick up Contestants at Willard Hotel
5:00 AM	Breakfast at National Airport
5:45 AM	Weather Briefing, Flight Plans and Departure Procedures by CAA.
7:00 AM	TAKE OFF WITH BENNY GRIFFIN AS OFFICIAL STARTER.

The next morning the fog in D.C. was too thick to take off, but they got us up at 5:30 anyway to eat green scrambled eggs. Drawing a deep breath before starting to eat, I never imagined the laughter I would share at the start of almost every race over a breakfast of eggs in that same hue. I came to think it was a race requirement to have a stomach strong enough to start the day with them. To this day the memory of those eggs seems to have permanently destroyed my ability to eat any more scrambled ones.

When we went out to the airport the next morning, we couldn't see a thing. Back to the hotel, green eggs, then back to the airport. The weather didn't look any better. In fact, it looked worse, but they said, "Pilots, man your airplanes."

Joan and I looked at each other with one thought (Ok, we'll go through this little show) and ran out to our plane to settle into our seats, secure on their tracks. We still really weren't planning on racing that day, but all of a sudden, "You have one minute 'til take off." Well! The flag dropped, I shoved the throttle in, my seat slid back, Joan's slid back, the charts flew even further back and we were halfway through Virginia before we got the cockpit re-organized.

NINETY NINES, INC
INTERNATIONAL AIR RACE

WASHINGTON, D.C.   June 9th 1955 - June 11th 1955   HAVANA, CUBA

Sponsored by:

FLORIDA CHAPTER 99'S. . . . . CUBAN TOURIST COMMISSION

****

AIRPORT	CITY and STATE
Byrd Field	Richmond, Virginia
Rocky Mt Municipal	Rocky Mt., North Carolina
Raliegh-Durham Airport	Raleigh, North Carolina
New Hanover County Airport	Wilmington, North Carolina
Florence Airport	Florence, South Carolina
Charleston Municipal	Charleston, South Carolina
Travis Airport	Savannah, Georgia
Brunswick Airport	Brunswick, Georgia

(FLY SO LOW THAT THE TOWER CAN IDENTIFY YOU) (VISUAL CHECK)

AIRPORT	CITY and STATE
Jacksonville Municipal	Jacksonville, Florida
Daytona Beach Airport	Daytona Beach, Florida
New Smyrna Beach Airport	New Smyrna Beach, Florida

(FREE LODGING IF YOU STAY IN NEW SMYRNA BEACH) (CLOSE YOUR FLIGHT PLAN IF YOU DO STAY OVERNIGHT)

Melbourne Eau Gallie Airport	Melbourne, Florida
West Palm Beach International	West Palm Beach, Florida
Broward International Airport	Fort Lauderdale, Florida

(A COMPULSORY STOP - ALSO CALL MIAMI TO CLOSE FLIGHT PLAN)

Miami Airport (ONLY TO CALL IN AND CLOSE FLIGHT PLAN) Miami, Florida
Meacham Field   Key West, Florida
CUBA

THEN HAPPY LANDING AT HAVANA, CUBA

After reporting our checkpoint for Savannah, Georgia, we were flying so low we scared a man on a golf course. Then we noticed the clouds of a big old frontal system across the only way we could efficiently get to Florida. In the International Race, it was legal to fly "on top" provided you could get above the clouds using Visual Flight Rules (VFR) and also back down flying through a hole in them VFR, so we flew up to watch gorgeous towering cumulus and zapped on down to Key West.

It was just about sunset as we came into Key West for our landing. The rules of the International were that we had to be on the ground by sunset. As B talked to the tower, I put on the landing light to help them identify our plane. But when it came on, a whole stream of cars headed for the airport. The people of Key West saw our light and thought an airplane was in trouble, so they wanted to see the excitement. Even the mayor!

Husbands were at the airport to meet their racing wives, but only two planes arrived as expected, and neither held any of the wives. For two or three days we burned ourselves in the sun while the weather further north settled enough for the other pilots to catch up to us. At first in the evening we went out with the mayor's entourage. They said he was trying to check out the terrible dens of iniquity around the naval base, but soon we called it quits because we realized he was using us as a shield; he already knew where all the girlie shows were.

At the final cocktail party before the race heated up again, a group from the Cuban Tourist Commission came over and introduced themselves as a goodwill gesture. We suddenly realized we were going to a country where we didn't know the language. With some shock we also learned we would have trouble pronouncing names correctly, much less remembering them. Still, in the morning, we all tore off for Cuba.

Date	Point of Departure	Time Off	Point of Landing	Time of Landing	Time Enroute	Accumulated Time
5-10	Washington National	9:29 std	RDU	10:52 std	1:23	
5-10	RDU	11:30 std	SAV	13:29 std	1:59	3:22
5-10	SAV	14:06 std	SSI	14:35.5 std	Visual check	
5-10	SSI	——	DAB	15:38.5 std	1:32.5	4:54.5
5-10	DAB	16:52 std	FT. ENDED	18:29 std	1:28	6:22.5
5-10	FT. ENDED	18:38 std	EYW	19:34.52	:56.52	7:19.42

I(We) hereby certify that, to the best of my (our) knowledge and ability, the above record is true and correct.

Pilot's signature_____
Co-pilot's signature_____
Date_____

According to our draft of the official flight record, Joan and I landed at Raleigh-Durham, Travis, New Smyrna Beach and Daytona Beach Airports on our way to Meacham Field [SSI designation is now used for an airport in Oklahoma].

FLYING ACROSS THE OCEAN

I don't remember being afraid crossing the ocean for the first time, but I have to admit that when the Coast Guard came to explain what would happen to us if we had to ditch and how we could best survive such a catastrophe, it focused my attention on the details. I knew that successfully ditching any aircraft was difficult. I was reminded that a plane hitting the water, gear down, will flip over. We had a fixed-gear plane!

The Race Committee required Mae West lifejackets, so we rented them. These unwieldy things are appropriately named when inflated, but we didn't even have to put them on in the flattened-vest form during the race unless we suddenly had an emergency.

With all the extra preparation, we were a bit more apprehensive as we left the coast for the ninety-mile crossing than when we arrived in Key West, so on this first race over the ocean, I gained more altitude than I normally used when racing across relatively flat terrain. I think I was trying to get high enough to keep land in sight, but, of course, that didn't work. Today, with more people flying across various parts of the ocean, I understand even better the ramifications of ditching a plane and I wonder at my lack of concern the first time over all that water.

However, we quickly settled in to focus on the instruments and listened intently to the radio chatter, keeping track of where everyone was. I don't know what we thought we might be able to do if someone got in trouble over the ocean, but it kept our minds occupied. It was a clear day, so even when we were out of sight of land, we did have the reassurance of a horizon, straight as a ruler in a complete and even three-hundred-and-sixty-degree circle with our little Cessna 180 in the center of it. We were released for the flight from Key West in such a pattern that, at our speed, we did not have any other plane in view. The flight didn't take very long, but it did seem as though the glance out the window was nothing but water for a long time. We were delighted when we saw the Cuban coastline rise up out of all that ocean.

At the post-race party Blanche Noyes asked, "You didn't go over seventy-five percent power, did you?" She was such a square, but she was highly respected in aviation. I wasn't sure whether to take her seriously or not because I thought everybody flew at one hundred percent power, just as we did, so I didn't answer her directly. In any case, it was fun to have Blanche take an interest in me. She knew everybody at the race quite well, including individual WASPs and she introduced me to some of them. It was a thrill to meet so many pilots I had admired, read about or had seen on the newsreels.

We didn't learn until a couple of days later at the banquet that Joan and I had won! To our surprise, they gave us a four-foot Batista Trophy with a little airplane on top, as well as a number of other trophies for speed and category challenges that we met better than other pilots during this race. After standing around with our arms full of trophies, beaming

Look at all those trophies! (The big one stayed in Cuba. We have replicas.)

Photo: Cuban Tourist Commission

for the cameras, they told us there was a limousine ready to take contestants to our hotel. Off into the night we went.

We weren't sure what direction we were supposed to be going in, but it seemed we were taking longer to get back to the hotel than it had taken to get to the banquet. Finally, we decided we needed to try to tell our driver to take us straight back to our hotel. He kept rattling off something we mostly couldn't understand, but eventually we began to realize he was taking us to a nightspot, because, "Everybody goes to celebrate and dance in Cuba!" He also told us in pretty good English that he had driven for Al Capone. So we settled back until he pulled up at the Tropicana, and the rest of the group got out and went into the building. Joan and I were the youngest contestants. We didn't want to follow the rest of the crowd. We also thought the driver would take us back to the hotel. Instead, he drove to the parking area, got out and left us. The lot was dark. It didn't take long for us to realize that it was no place for us to wait, so we got out, sacrificing the trophies if need be.

As we quickly walked from out of the darkness into the lights, some people in what looked like a ticket booth kept motioning us to come on, so we went over, bought two tickets and went in the front door. The maitre-d' escorted us to a nice table for two. When I placed our tickets on the table, the man who seated us took one look at them, looked hard at us, rattled off something and left. He came back with a man with a black jacket and white pants who began to speak even more rapidly, creating a jabber of sound. We couldn't understand anything, which made both men talk louder. They finally went away and brought back a man with a white jacket and black pants. These three men then jabbered back and forth, ignoring us. We still couldn't understand what the problem was, but we noticed that by this time nobody was watching the floorshow. We were so embarrassed! Finally a man with a white jacket and white pants came. He began talking louder than any of them until we couldn't hear anything at all above his noise. Finally, out of the crowd came one of the men we had met in Key West. He asked if we were having some difficulties. I tried to keep a straight face. "Apparently we are. I don't understand what the problem is, but we haven't been offered a drink yet." He looked down and saw the tickets. "What the hell! Where did you get those?" "Outside at the ticket office." Our suddenly quite definite friend, the Cuban Tourist Commissioner, jabbered away at the collection of suits until they all left.

"Would you like to dance?" "Ah, I don't know." "Come on, relax." We went out on the floor. "I just wanted to get you away to explain. Down here in Cuba, prostitution is legal, but you have to buy a ticket. You bought the tickets, but the rule at the Tropicana is that prostitutes are not allowed on the floor. They have to stay up at the bar." Joan and I were... !

We may have been young women and fresh from a different culture in the United States, but we already knew enough to guess that being an American woman in a Cuban jail would not be appetizing. Tabbed as prostitutes, all we wanted was to go back to the hotel and tell the world that we had won.

The next morning when all the fuss was over and we prepared to fly home, I suddenly remembered Zaddie Bunker, a new acquaintance, who had been overdue two days at the end of the race. The response of the racers to her plight was typical of the Ninety-Nines. As tired as we were from the successful end of the race, we all went back up in our planes to try to find her. Zaddie was just about to land in a muddy field in the middle of Cuba when Fran Bera talked her over to the airport.

As we finished getting ready to fly home, I worried about Zaddie getting lost again so I went over to that dear old lady and asked her to fly with us as far as Raleigh, Virginia. Then when we got to Raleigh and were about to continue the flight to Ohio I asked, "Are you all set? Do you have enough money?" "Well, Dearie, if the Chase Manhattan hasn't gone out of business, I do."

Zaddie was one of their leading stockholders!

FURTHER ADVENTURES OF JOAN AND MISS B

I started to pay even more attention to Joan as a pilot when she came to a Paul Bunyan gathering in Traverse City.

The Paul Bunyan Clan predates WWII. B and her husband reactivated it to make the Michigan Sectional Meetings stand out. Inductees actually wind up kissing a live ox, Babe of course.

The list of people who have done so is rather impressive! The Clan was developed among pilots when airtours were popular before the war. It involved the voluntary efforts of a lot of the local community. Many of the original Paul Bunyan people were still active in the flying crowd when I moved to Traverse City with my husband. We benefited from their continued interest, support and excitement with our reactivation of the Clan. A wonderful, living Babe the Blue Ox was located, a full-scale indoctrination into the Clan was developed, Babe was groomed for a number of kisses, and we got all dressed up. The anticipation of an extra bit of fun attracted more people to the sectional meeting held in Traverse City than ever before because the families of the pilots were invited to join the Clan and to participate in other activities while they were in town.

Janey Hart's F-model Bonanza. Notice the stork.

Joan also flew the 1957 AWIAR with me in Jane Hart's F-model Bonanza. We didn't win that time, but the race had its humorous moments. Picture us in Jane Hart's Bonanza with a stork painted on the side. We crossed the finish line so fast we had gas streaming off the trailing edge just like a vapor trail.

Well, at least we landed in Cuba very much in a hurry, and still a bit disorganized. We were warned before this race to behave like ladies; we were to wear skirts over our shorts. We suddenly realized that in the heat and confusion of ending the race we hadn't put our skirts on yet. There we were, screaming down the runway, trying to get our skirts on. Joan is about five feet tall and I'm about five feet eight inches. We figured out we had the wrong skirts on just about the time these darling Cuban men came up talking Spanish again. I still couldn't understand them, and I couldn't make them understand that we weren't ready to get out of the plane yet. Oh yes, it was a speed race.

We found a change in Cuba on this trip. This was the first time we heard some discussion of Castro. The 1955 race ended at a small airport outside Havana. From the airport we were taken to a small café in the old slave quarters and served coffee and a huge plate of something delicious that looked like donut holes, but were made with a heavenly dough. This time the island was under martial law.

This was also the race where Joan's parents met us in Florida. In addition to the required Mae Wests, which we rented again, her folks asked that we carry a raft in the plane because Joan couldn't swim. We wedged that raft in the back so tightly that, at the end of the race, they nearly had to dismantle the plane to remove it. A lot of good it did…added weight, too.

Our friendship also led us into other small adventures. One time when my folks weren't home the two of us tried to make daiquiris the way they made them in Cuba. We never did get them right, but we were schnockered by the time my folks returned.

Better adventures came later. When I was recovering from major surgery, Joan really proved her friendship. She drove around the Upper Peninsula of Michigan (U.P.) with me in a little Ford, looking for prehistoric copper mines. Sometimes we didn't know where we were. We used an old compass to follow our way out the old logging roads. One time when we returned home, the car broke down the very next day!

Yes, but before the time in the Ford, Jean Reynolds, Jackie Scott and I went to the U.P. with B in Bertha, the Winnebago. Bertha's water tank leaked like a sieve. So did Bertha when it rained. We entered Canada through Sault Sainte Marie on the way up and returned to the U.S. via Port Huron.

Bertha's bumper broke off as we crossed the Blue Water Bridge. Everyone at the truck stop we pulled into was fascinated by a Winnebago full of women. With all those marvelous Canadian china and woolen shops, all of us were broke by that time. Fortunately, the repair shop didn't charge us for the repair.

Vehicles always have a way of being a part of our story. After B married Bob, I helped them take their thirty-six foot Marblehead cruiser through Lake Huron and Lake Michigan as far as Traverse City. We had to pull into Cheboygan for repairs. When we finally got into Grand Traverse Bay, the marine repairman told us he couldn't believe we'd made our trip successfully!

Later B owned an old funeral car, a Cadillac limo. She set her first son's playpen up in the back, where it had its own heat and air-conditioning and off we'd go!

Then there was the time Jackie Scott and I went down to help Joan take care of something for her mother. Joan impressed Jackie with her sense of calm when my Peugeot broke down for the last time in the middle of traffic. All kinds of people had to help us then as we each had to go in a different direction to get back to our jobs. Jackie says I declared, "They should sell each of these cars with its own tow truck!" But I loved as well as I hated that little car.

The Small Race

The race that later was called the Southern Michigan All Ladies Lark (the Michigan SMALL Race) was developed for the end of the tenth annual transcontinental race in 1956 while I was on the Ninety-Nine's transcontinental race board. Our national convention was going to be in Harbor Springs, Michigan, after the AWTAR, with my friend Mary Clark in charge. When I flew into New York for the '55 Convention and Race Board meetings, I asked them if they would end the race in Flint if I could get a good sponsor. When they agreed, I returned to Michigan and set everything in motion.

As I worked for various flying services I continued to charter some General Motors-A. C. Spark Plug Division executives and did a lot of commercial hauling for them. I still kept in touch with some of the people I'd met during the flight testing for AC's gun sights. I knew that some combination of these people could do a gorgeous job of rolling out the red carpet by sponsoring a terminus, and they did not fail me. They put up the money the AWTAR board required prior to the race, arranged a big cocktail party, acted as hosts, furnished transportation, and helped with printing and hospitality. It was wonderful. The people AC assigned to help worked well together. I could simply focus on other aspects of the race.

Mary was in charge of things in Harbor Springs. If I could get the transcontinental pilots to Flint, how could she get them to the convention? We knew we were going to have a successful terminus for the AWTAR, but people who race and people who come to convention are often not quite the same group. We wanted to bring them all together. My reliable Mary and I began to plan together. Margaret Crane, Jean Reynolds, Helen Wetherill, Leah Higgins... I can't remember how many were there, but a group of Michigan Ninety-Nines joined us one winter night at Margaret's house. With a fire in the fireplace, good conversation, much laughing and lots of coffee, we decided the best way to get the racing pilots to Harbor Springs would be to have a race.

We figured the best run race was the AWTAR so we took their rules and regulations, with Ninety-Nine permission, and adapted them for our race. Because it was not going to be a long race, we called it the Michigan Small Race. It was not until some years later, as the pilots were all waiting around for the weather to clear, that a contest was held to figure out what "small" should stand for. That contest was a riot. Many of the word groupings were humorous. One that I remember was "some men make lousy lovers". There were at least twenty different suggestions, but when the laughter settled, the name we selected was the Southern Michigan All Ladies Lark: (SMALL as opposed to the BIG transcontinental race).

Since all the transcontinental racers were coming in fast, ours became an efficiency race. The idea turned out to be a lovely flyer. We had more racers at convention than ever before, even though the weather up north was not that great.

Nineteen Fifty-Six

The year 1956 turned out to be very difficult for me. It started on a high note because I was excited about the race, but then my fiancée, Dale Hath, was killed. Pellston, the largest, closest airport to Harbor Springs, has high hills to the south and west. When the husband of one of the racers hit the top of one of those hills with his airplane and was killed, it intensified the shadow on my soul. I had to tell his wife about it as she came in from the race. That was a bad day.

In addition to all this, the man who replaced Dale at Flint Aeronautical was a pain in the neck. Really, he was just absolutely unbearable. He insisted I work every day for him in Flint, even when I was developing a national event that could increase his business! It seemed to me that it was more important that we carry this whole event off successfully, but I had to work on the race despite him. As the time of the convention approached, I flew north to take care of things, then flew back so I could work at the airport. I don't think many people were aware of what was going on in my personal life then, but AC was aware. They offered me rides back and forth with their people in a company airplane. I think I flew back and forth two days by myself and rode back and forth two days with them. It was a tough challenge to balance everything that year. I did not race.

Fran Bera was the winner of the '56 AWTAR. Some of the other racers were: Dr. Dora Dougherty, in charge of human factors for Bell Helicopter; Edna Gardner Whyte, whose racing days are memorialized in *Rising Above It*; Mrs. Graham-Bell; and Eloise Smith, operator of the airport at Austin Lake in Kalamazoo.

I have photos of "The Social Lights and the Pi-lights". This skit was all about flying and was performed during the banquet at the end of the AWTAR. The laughter it generated helped get AWTAR pilots focused on Harbor Springs. Other women kept track of where all the planes were every day during both of the races.

Ann Fruehauf, with whom I would soon fly to Florida, Lucille Quamby and Babe Ruth participated in the first SMALL race. Betty Miller, who went on to become the first woman to successfully fly the Pacific, also raced up to Harbor Springs. Betty and I were talking about flying the Pacific together, but when she said she was going to use her Apache, I did not go. I wouldn't have used my Apache even to fly Lake Michigan. Eventually she found a better plane to use for her flight, but by then, I was otherwise occupied.

We consulted Piper, Beech and Cessna before the race, asking them to give us handicaps for seventy-five percent normal performance with fuel consumption at sea level so everybody supposedly would have an equal opportunity to win. We had graphs for the various models that each company manufactured. Beech didn't like this idea. They said point blank, "We don't sell airplanes for fuel economy; our airplanes are fast." Nevertheless, we got their cooperation. We established handicaps for all the entries we expected to fly in the race.

Alice Roberts, a pilot from Arizona, won the first Michigan Small Race. I won the second, flying between Lansing and Traverse City.

SMALL RACE CONTINUED

The SMALL Race has continued as an annual event since that AWTAR. We don't want the racers to get the idea they know how to beat the system of scoring, so we change the method of scoring from time to time. One year it was who used the least amount of fuel. That challenge got started when gas was rationed: how fast could we go on how little fuel.

I flew a Cessna 190 with swivel gear called "Crazy Legs" in one

Michigan race. My co-pilot, Lucille Quamby, didn't like the plane. It is hard to see around the cowl of a radial engine. The main gear of Crazy Legs swiveled so you could taxi at an angle, with the nose pointed to one side. Not all C-190s had that gear. The oil additive I used before the race to clean the engine plugged the fuel strainer. Our oil temperature shot up: "look for an airport." We did get to our destination but it was a close thing. Later, when Crazy Legs had several forced landings, it got a new engine.

Another year I raced in a Bonanza, burning something less than eight gallons an hour at an average of one hundred twenty-seven miles per hour. We earned points for both fuel economy and speed. They called the factory to make sure such efficiency was even possible! This kind of precision still attracts me.

The Michigan SMALL Race also creates an incentive to upgrade piloting. One year we decided each PIC had to have a commercial license and a licensed co-pilot. Everything is set up to draw attention to small airports and to get their communities involved. An airport brings dollars into their economy. We celebrate a new airport, or a new FBO on a field, or even a new runway. When this race began, we flew to a destination and stayed overnight. Both the starting and terminus points would put up money, so we would have prizes. Then we decided to fly a triangular course, approximately three hundred miles. This round-robin started in 1960 and became a navigation race. One year we were especially glad for this change. Bad weather lasted for so long, we held the banquet before the race. Never say pilots are not adaptable!

After 1956, our SMALL Race became very popular. Other chapters have since had small races, asking to use our rules and regulations and to use our process of determining the winner. There are still several across the country, from California to the East Coast. In the beginning they all patterned their races on ours, then developed their own ideas. I'm sure ours was not the first short race in the country, because out in California they were doing similar things, but I believe ours was the first all-women event.

Now even our SMALL Race is co-educational, and is officially called the Michigan Small Rally. But when the planes are lined up and numbered in position on the flight line for the start of the Rally, position number one is still reserved for Sammy McKaye, to honor her as the

Thursday, Nov. 28, 1957   CROSS COUNTRY NEWS

The S.M.A.L.L. Race

By ELSIE FERICH

You could almost put your finger on the excitement and tension that was in the air at the Pilot's Briefing the evening before the Michigan S.M.A.L.L. RACE.

October 4 was a sunny race day, and the answer to all prayers. Winds were reported 20 to 25 MPH and gusty, but it was CAVU all 176 miles of the race route. We made our turn on course and race or not, the hand of Indian Summer had left one of the most breathtaking sights to see. It was a panorama of vivid siren colors.

I stayed low to save time and gas rather than climb to winds that could do no good. My copolit reported ground speed 88 on the first check, then 90, and 93. Just before reaching our first dog-leg, she gave a ground speed of 98, and our spirits soared. We stood a good chance if we could just hold that. Gas consumption didn't worry me, having throttled back to 2050 from the 2150 I normally use, thinking I'd save gas.

We dropped to 250 feet MSL, and buzzed the State Police Post at Houghton Lake and were thrilled doing it. It was cleared and legal. We headed north to Grayling, and zoomed past the control tower, our second dog-leg, then swung west on the final lap. My ground speed had dropped, and the N NE wind I had been fighting, switched like a fickle woman, and was again on my nose. The remaining gas supply was disappointing . . . it was down much more than I expected, but as I buzzed the finish line across the Traverse City Bay, I felt like a winner!

Joining the traffic pattern, I came in with a power-off landing, letting the T-Craft float leisurely down the long runway. We taxied to a parking position and directed by the official, cut the engine. We had made it!

The race results for the fifteen ships were compiled by that evening, and the Post Pilots Meeting was held. Congratulations were in order for, Bernice Trimble, Sammy McKay, Laurien Griffin, Jean Bonar, and Alice Hammond. These girls are the greatest, and seeing them in the winners circle makes you feel like you won too.

I was notified I was T.E.T. Me, the kind of person who never wins any prize, first, last or booby, had gotten a trophy. I'm proud of it. It represents a start, and an incentive to race again. I know when the flag drops, Tail End Me will be there with full throttle!

person who has flown the greatest number of times in the race. Sammy also won the race three times.

44th Michigan SMALL Rally
(Formerly known as the Michigan Small Race)

September 15-17, 2000
Official Rally Course

Leg 1	Cherry Capital, Traverse City to Clare Municipal Airport (48D)	65.5 nm
Leg 2	Clare Municipal Airport (48D) to Evart Municipal Airport (9C8)	23.5 nm
Leg 3	Evart Municipal Airport (9C8) to Empire (Y87)	62.5 nm
TOTAL MILES...		151.5 nm

Empire (Y87) to Cherry Capital Airport, Traverse City......19.2 nm

A More Complicated Air

(with Bob and Berniece Vaillancourt)

Bob Vaillancourt and I did not stop helping each other as he began to move up through the ranks at GM Transport. He was really a good contact for me there for many years, someone who reinforced what I was trying to do by myself.

(Bob V) As B went on with her career, Berniece was devoted to our children and this old line boy buckled down to the aviation maintenance program at the Buick hangar in Flint. When they thought I needed to fly co-pilot for one of their crews, I took my commercial in a great hurry with B's help. Then I had the fun of flying a larger, more complex airplane, but I was never sorry to go back to maintenance. It was a bit more than an eight to five routine. In those days a corporate pilot who flew to the west coast stayed a week. Today you might fly different people different places every day, but spend your relief time and evenings at home with your family.

But unlike B, I never raced an airplane. I thought that was too wild. Being in maintenance, I think any squealing tires go against my grain. It's the same thing with an aircraft engine. If I flew with somebody running full throttle for four hours at a time, I'd probably throw up.

I think that's right. Different people have a different focus. One time I called the Lycoming plant and told their engineers, "I'm going to pick up an airplane with a new engine, and I'm going to race it right away." They said, "That's all right, we guarantee it for one-hundred-fifty hours wide open." This made me an great fan of Lycoming engines.

Racing Turns

I began to go more my own way when I started pylon racing. Berniece wasn't interested in pylon racing and, as he said, Bob Vaillancourt couldn't tolerate stressing the airplanes. I was racing against other instructors from competitive flight services located on Bishop Airport, with different kinds of airplanes. I remember one instructor had a Luscombe, one had a Piper J-3 Cub, another had a Monocoupe...I'd get an old Taylorcraft out.

We flew around pylons in our local area and, happily, the races were co-ed. I was on my way to getting a commercial license. There was a tremendous difference in the planes we were racing around the circuit, but an even greater difference in the performance abilities of the pilots. The sixty-five-horse T'craft did pretty well, but the Luscombe was fastest. I couldn't beat the Luscombe on the straightaway, but I sure could in the turns. Curved flight causes acceleration, which makes an airplane climb. I didn't want to waste this added speed in a gain of altitude, so it was a delicate challenge. If we lost speed by climbing in a turn about the pylon, other pilots went zooming past as we returned to the straight part of the course. How high off the ground were we? Legally? They wouldn't want me to say how high we were flying. We were pretty darn close to the earth. We would pull three G's (a force that made us feel three times our normal weight) on a pylon course that wasn't very long; between three and five miles.

The pylons were selected geographic points: corners in the road, barns, anything that stood out on the landscape. The competition we set up was a tremendous inspiration for each of us. We were trying to shave tenths of a second as we flew around those pylons. I think we hangar-flew ten hours for every hour in the air. Pylon racing is another thing you simply can't go out to do with all the regulation and liability issues of today. Back then, it prepared me for bigger and better challenges.

One Saturday somebody slit the fabric on my plane. I got very melodramatic. "They know I'm going to win, so they're just making sure I don't beat them." We used a simple adhesive tape to repair the slit so it wouldn't tear any more, and went on racing. Was that courageous or was that foolish? I'm glad it wasn't too dumb.

Promoted

When B was racing pylons and I was working for GM, they probably had the largest corporate fleet of any business in the world. When I became Super, the president of the corporation and the chairman of the board might be away in Washington, D.C., with one of our sophisticated airplanes. I'd get a call saying they were getting ready to depart, but something was not working prop-

erly on the airplane. "What do we do?" I had to give them some fast answers: send them alternate transportation, or fly out and fix the part.

Today's environment in maintenance is almost all remove and replace. Now you rarely do more than pull one black box out and put a new one in its place. The new method has become more cost-effective because of comparative prices. The challenge is in deciding which area in the vastly more complex airplane has the problem. If your corporation has a ten million dollar airplane sitting on the ground, no-one worries too much about whether it's going to cost two or three hundred dollars versus a thousand dollars to get it back in the air. They just switch units and go. They can change compressors, change turbines, pull and change whatever the plane needs. Then we send the problem units to a vendor. The vendor overhauls them and sends them back so they can be reinstalled in the future. This can continue until the scope or some other process shows metal fatigue or another form of too much wear and the unit is terminated.

In the old days of piston aircraft at Buick, we tore the engine right down when there was a problem. We would even weld things. If a wooden spar broke, we would splice it with a new piece. Sometimes we spent two or three days making repairs on one airplane.

In contrast with the work on the corporate planes, we spent very little time on the maintenance of the Taylorcrafts we rented to B when she opened her business. I didn't fly those planes. They were only for the business. I mostly did oil changes. B flew the T'crafts so much, we changed a lot of oil and spark plugs. About the only thing we might have had to pull to replace on our old T'crafts was a bad cylinder. Fortunately, with those particular planes, I don't think ever we had any of that to do. We did have prop replacements once in a while, however. I well remember one.

That was when they told me not to marry one of my students. Good thing I ignored them.

Changes in The Wind

In addition to the planes we rented to B, Berniece and I used to buy other T'craft, fix them up and sell them. We spent all the early years of our marriage recovering airplanes while B was buying up MGs and selling them.

The only plane I flew for anything like recreation then was an 1946 Ercoupe. Having the right to use such an airplane was just about like owning it. Unlike owning it for real, however, it never cost me anything to fly. I flew

The Ercoupe had a simple panel.

it a couple of hours a week. I was to keep the rust out of the engine for Mr. Wiles. When I'd get done flying, I'd pull up to the gas pump, fill the tanks and sign for the fuel.

The real owner of the plane, Ivan Wiles, was a GM VIP. He was then general manager of Buick, a real fine gentleman. We maintained his plane at the Buick hangar. He had a vacation place up in Harbor Springs, Michigan. If the weather turned sour, he would leave the Ercoupe there and drive home. Then I'd get a call to go up to fly it back. The rest of the maintenance crowd loved me for that, including my own boss. He didn't think taking off to fly in someone's private plane was good practice, but I enjoyed it! I also used it to fly into Detroit City Airport to pick up Mr. Wiles. There I'd get into all kinds of trouble with the Director of Air Transport.

"You come down here with that little puddle-jumper to pick up an executive of the corporation.... . We won't go for that at all. You're not employed to do this!" He had a fleet of airplanes out on the line that he maintained to fly the corporate executives. It seemed like I was a threat to his image with a little eighty-five-horse Ercoupe. Finally I said, "Don't talk to me! Talk to Mr. Wiles. I just do what he tells me to do." Later, when we got transferred to Detroit, this director met me with, "Remember when I had no control? Today things are different!" And they sure were.

Detroit City

I worked in Flint seven or eight years before they closed the Buick wing of the GM Flight Department. Then they took our Lockheed PV-1 and Lockheed Lodestar to Detroit. Out of work in Flint, Berniece and I followed the work to GM at Detroit City Airport. Later they moved to Willow Run, then on to Metro. I retired in '87 and left Detroit and moved to live on this golf course in the middle of the state. GM is still at Metro.

The A&P maintenance crews had good times in those early years in Flint. We harassed the pilots, playing games. When I first went to work, we used to laugh a lot. Labor was cheaper in those days. Now days everything is so expensive, you can't have anyone sitting around planning mischief. Toward the end of my working life in Detroit everything was cut and dried.

OSHA (Occupational Standards and Health Authority) came into the picture. Many of the pranks we'd pull on the job in the early days of aviation we wouldn't dare do now. We could help people more before they began to regulate everything. If somebody needed a bolt for an airplane, we could give them a bolt. Now they'd say the bolt was defective, so the wings fell off. No matter how the pilot or some weather stressed the plane out of the rating envelope, if it fails, it seems to become the former mechanic and owner's liability. General aviation industry in the United States was in such trouble with all this liability insurance for a while, they almost went out of business. We fell behind the rest of the world then.

At one time the Experimental Aircraft Association (EAA) advised builders not to sell their experimental airplane. Give it away, put it in a museum. The person who built or modified an airplane seems to be responsible for the life of future pilots of that plane no matter what the actual circumstances of an accident might be.

Though we stopped being in business together, my life remained interwoven with the Vaillancourts after Bob & Berniece moved to Detroit. I'm glad we have remained friends through challenge and change. Like the proverbial good wine, our friendship just gets better with time. Now we play a mean round of golf together.

Ties That Bind

(Berniece) *I couldn't afford to fly after we were married...for a while. After I had children, B wanted me to go to the meeting of international pilots, the Ninety-Nines, in Kansas City. We joined that marvelous group of women shortly after we got our licenses. I wanted to go have some aviation fun so much! Mamma sent the priest over to talk to me. Did I want to leave those children motherless and so forth and so on? I cried... .*

I didn't go. I completely quit flying until we bought shares in a Cherokee

180. Then I simply went! When my mother was over seventy years old, she started to fly commercially. I was really angry with her, yet proud of her, too!

All Bob's talk about mechanics reminds me that B also has a passion and an unusual eye for automobiles.

Well, in a sense, Berniece started that!

Yes, working for GM, my dad would drive around in special-transmission prototype cars. Remember the Dynaflow? That car would go from a dead stop so fast!

I was pretty bad. When I finally got a license to drive I'd sit at a stop sign and look at the guys in the car next to me. Once those fellows had a good look I would go off. Whewwwwww! That thing was jet-propelled. My dad would ask, "Now, you aren't using that special gear are you?" Was he serious? With it, I could beat anything.

When Berniece drove around scaring the boys with the Dynaflow, it was an experimental; the first car with an automatic transmission. My own first car was a Pontiac that was pretty outstanding. It had special leather inside, all red and white.

From then on B always had fancy cars. One day she came over, feeling sorry for me, stuck with two kids in diapers who seemed to be about three months apart. "Can you go for a ride with me?" I'd been feeling pretty unattractive. Out with B I saw guys looking at us and I felt, well, maybe I didn't look so bad. Finally I figured out they were looking at B's car, but it still didn't matter. "What kind of a car is this?"

It was my first Mercedes.

You eventually got so many tickets you had to quit getting those beautiful cars.

That was only in a Jaguar. I do have to admit that Jag got quite a few tickets. It finally drove me to my day in court. The Justice of the Peace could not believe I was driving as fast as the ticket said I was. He also thought I should have talked the officer out of the ticket, but finally he said, "I'm going to take your license away for six months!"

"Oh, you can't do that!" "Why can't I?" "I've got to get to the airport to work! Otherwise the transportation I have is by bus or by bike." I also exaggerated the distance I would have to walk and, by promising to behave, I got off. Then, somehow, I faked good behavior well enough to avoid further trouble. I guess I finally realized such speed was not meant for the roads I was driving and, perhaps more important, that I couldn't resist the speed.

I was never really in the business of buying and selling cars, but I did have a penchant for nice cars, sporty cars. After the Pontiac, I had the cutest little Chevy coupe with a hat shelf instead of a back seat. A lady who also owned an exclusive ladies' specialty shop was the first owner of the car. It was a medium blue with gray mohair seats.

Miss B and her jag. (Air Force photo)

But the real car fancy started with my first MG's. Eventually I owned an Alpha Romeo Julietta Spider, an Austin Healy, the Jaguar, and then the Mercedes, each because I liked the workmanship. It was not that those cars were all magic. They certainly were not! They constantly needed repair and it was hard to find qualified mechanics.

My cars did not relate to going into business for myself, but they made me appear to be successful. I look back at them and wonder how I managed to own such cars on the money I was making. But back then I could find cars at a good price, buy them, fix them up, sell at a profit and buy something else. It was nothing for me to have a couple of different cars a year. I don't remember losing any money except with the Jag. If I had kept that Jag, I could retire now by selling it, because it was a convertible, a rare bird, a truly gorgeous thing. But it was the most contrary machine I have ever had...no, the Peugeot was the most ornery car I've owned.

Those cars reinforced my knowledge of the mechanical systems we had to understand as we took both the written and flight tests to advance in aviation. When I buy a car now, I still just look at the nice ones. Last year I went through three, ending up with a Lincoln Towncar, a very smooth ride for traveling to aviation meetings.

I Bite the Bullet

I taught ground school at the junior college in Flint for the Mott Foundation when I was working for Charlie Glidden. I was told at the time that this adult education class was the first such in Michigan. One of my students, Mott Lyman, had a hardware store in Montrose and became a good friend. "If you're going to work this hard, flying all day and teaching at night to make money for somebody else, you should be doing it for yourself." It was a nice idea, but I didn't have any money. I was still living with my parents. I owned a sports car, bought my own clothes and had what I thought were all the necessities and most of the luxuries. But I did not accumulate a great deal of money. I didn't see how I could start a business. Mott listened to my excuses, then said, "I think you can do it."

So I began to study the question. In addition to the usual things you might expect a business to have, I found that to open a flying school I had to own an airplane. It didn't make any difference if I used it for flight instructing; I just had to own it. As I was trying to solve this problem Bob Vaillancourt said, "Mr. Wiles' Ercoupe is for sale. It is in *beautiful* condition." He went on to say the plane had never been stored outside. Mostly it was hangared with the General Motors corporate planes. "This plane is polished, a showpiece." I had the chance to buy it for three thousand dollars, a very low price.

For what seemed like the first time in my life Dad surprised me by offering money to help me buy the plane. He said he wanted to help me start "the best flight training school in Michigan." Owning this airplane made me legal. But I still didn't have anything that I would use for teaching. That beautiful Ercoupe was an investment, not for student wear and tear.

The Ercoupe also was designed to handle like a car. Its tricycle landing gear was connected to the yoke for ground control. In the air, the co-joined aileron-rudder system eliminated the need for rudder pedals and was controlled again, with the yoke. This made the 'Coupe almost impossible to spin. However, coming in for a landing with quartering cross-wind, the pilot cannot use cross-controls so the approach is flown with the nose pointed up to ninety degrees away from the runway! This approach is not for the faint of heart. With such unique systems the 'Coupe does not prepare students for flying conventional aircraft.

To solve this problem, Bob Vaillancourt convinced another GM mechanic to help him buy two Taylorcrafts. Bob also agreed to do the mechanical work for me. I agreed to pay so much an hour for the use of their airplanes. Their investment launched my business. I remembered the encouragement of my old instructor, Ralph Rose. He said that if I could fly a T'craft, I could fly anything. As an instructor I learned to trust those old birds. They were honest airplanes. If my students could fly in and out of a small grass strip in one of Bob's planes, they were proficient enough to sign off for cross country.

Next, I went down to Chevrolet salvage and found an olive drab Stowe-Davis desk and a chair. I paid about ten dollars for both of them. I fixed the chair and painted the desk with some paint that creates artificial wood graining. It looked pretty darn good! Then I talked the airport

Trimble Aviation, 10th Air Force upper right window

The new owner with her first airplane.

manager into letting me set up my desk in the lobby of the old terminal building where Capital Airlines used to be. When they moved, the airlines left their old davenports and a couple of chairs, so I had something for people to sit on. Finally, I put a lock on the dial of the telephone because there was no way to secure my office, hung a big sign outside that said "Trimble Aviation" and was in business! My office was located between the men's room, the ladies' room, and the stairway that went up to the Tenth Air Force schoolroom. I had people coming and going. There was no way not to succeed.

The Tenth Air Force

Those Air Force Reserve men were already well known to me. They were primarily a group of heavy bomber people. Quite a few from the Flint area were pilots and navigators. In the early fifties, the military put out a contract to keep their pilots upgraded and to maintain their familiarization with new developments in aviation. Dale Hath, my original boss at Flint Aeronautical, applied to teach these classes. The arrival of the books was my first surprise. "What are we doing now? I don't know anything about all these books." "This is what we are going to teach." "We are? Are you going to teach these things?" "No, Miss B, you are!" "I don't know anything about this material." "Well, just read ahead of your students!"

Believe me, I did… just barely ahead. The Tenth Air Force was to learn navigation, meteorology, codification, Shoran, Loran. I kept just a couple of chapters ahead of them. Our contract was for three nights a month over a period of ninety days. When I was scheduled to teach precision-approach radar, even the reading was a little over my head, so I called Selfridge Air Force Base and told them what I was doing. They were working with everything I needed. "Come on over. We'll show you how it works." I spent quite a bit of time there and actually had the chance to talk a flight in on radar. I came back home feeling there was a great bunch of guys at Selfridge.

Another great bunch was in my class. There were thirty officers; the lowest rank was a colonel, so they were all very experienced. We had a great deal of fun. Some of it would have shocked my mother. I still have a lot of friends as a result of teaching that course. This experience reinforced something I was gradually learning through my regular flight instructing: the best way to learn all about something is through teaching.

Miss B instructing the 10th Air Force (Air Force photo)

I told the reporter the truth at the end of a news clipping from June 3, 1956: "[B Trimble is the] Only woman anywhere serving as a flight-operations instructor for the Air Force Reserve. ...[She is to] bring them up-to-date on electronic navigation-an extremely technical subject...[while continuing in her position as] Flint Aeronautical flight manager." The report went on to quote Col. George P Stubbs...'She's one of the best instructors I've ever heard...as good as any man.' [I modestly replied,] 'I think I'm learning twice as much as they are.'"

FLINT FLYING CLUB AND MORE
There also was a flying club at Bishop Airport with fifteen to twenty students. It had been in existence for quite a while, but the State Department of Aeronautics suddenly required all clubs to have one instructor responsible for the entire group: checking members out in the planes, keeping them current, reviewing other safety factors. They asked me if I would be interested. You can be sure I was very happy to have so many students all at once.

I worked with my own students seven days a week. To have a successful business, I decided the airplanes each had to fly everyday we had VFR weather for a minimum of four hours a day. Spring and summer I'd

Miss B checks out her secretary.

be out there instructing more intensely than when I worked for someone else. The only way we could justify having a day off was if I resurrected the cross-country trips. I'd instruct Saturday morning and we'd leave in the afternoon. We'd fly wherever I decided we were going and have a bit of fun planned. Returning Sunday morning, I'd be available Sunday afternoon to instruct again. The students were really happy with this challenge, and I'd come back refreshed and raring to go. I felt if we could have fun, I could keep my students' attention.

There was one weekend flight when that wasn't quite true, however. I took my students to the narrow grass strip between the trees north of Hessel in the Upper Peninsula. One of the men on solo stretched the end of the runway and landed in the trees. There was minimal damage to the plane; they could just patch the wing. The pilot was not hurt. As we raced over to him, he was sitting on a rock smoking a cigarette right beside the leaking gas tank. His smoking almost did him in, not the brush with the trees. I was left with an extra ton of paperwork for the accident report that week.

Many of my students went on to do a lot of flying. Some bought their own airplanes. Some went with Zantop (a local freight carrier). The airlines weren't hiring too many at first, then some did get in with big carriers. Some even went into military flying.

One former student who also instructed for my business recently contacted me. I kept track of Bob Cuple up through the time he was flying an Argosy for Zantop. When the Argosy developed an engine problem, he landed that plane on the highway right under an overpass. The impact took both wings off but he wasn't hurt. In the flight manifest something rather explosive was listed on board, so he ran back to get the flammable stuff out of the airplane as fast as he could. They grounded him for a while, then took him back. Bob went on to greater things. He's as happy as a clam.

Another of "my kids" is flying a Gulfstream 4 today for some Arabian oil potentate. His wife flattered me recently by saying that her husband was happy that I gave his life direction.

Growing the Business

In 1958, as my business began growing, I did enough flying to earn the money to pay my mechanics for the Taylorcrafts, but I was still parking them outside, and that is not good for fabric aircraft. One of the new kids working at the airport told me that a man had a hangar for sale down at the corner of the field in which he stored his de Havilland Dove. He was moving to Pontiac.

Well, marvelous. I had such a terrible cold, I could hardly talk, but I had flown for this man on charter trips. He was a polo player. I flew him several times to Chicago. Once he had me land the Stinson right on the polo field. I not only landed him on that field, I taxied him right up so the wing was under the edge of the tent. He thought that was pretty neat. In addition, I flew co-pilot in his Dove on several flights.

"I would really like to buy your hangar." "Well, all right... ." He quoted me a price. "I can't afford that." "What can you afford?" I offered him what I thought I could handle. He thought my offer was pretty funny, but I must have been a novelty to him or something, because he finally agreed. "How are you going to pay for it?" "Well, I need a long term loan." "What about the interest?" "I would be willing to pay you five percent interest." He choked. "All right, draw it up." My boyfriend, the lawyer Bob Steadman, drew up the papers. I had that nice big forty-by-sixty foot hangar paid off in a pretty short time because I never liked debt.

Now I had a hangar, but I didn't have an office. The only way I could get all my airplanes inside was to stand the taildraggers on their noses on racks and slide them in. With the use of the racks, I also had room for people to pay to keep their planes in my hangar. I knew I could get all the airplanes inside, but the building had no heat. I also had to have maintenance in the hangar and an office so I could sit down to manage the business.

The FAA was going to build a new instrument landing system housed in steel and concrete to replace the original structure made of wood. All their bright orange painted wood was going to be burned. My mind went over and over the situation. (The FAA is going to burn perfectly good wood...really good wood...marvelous wood. There is no good reason to burn all that good wood, not when I need an office.) Somehow, Trimble Aviation got that wood and built a little office inside the hangar...somehow. It reminded me of the cornstalks we gathered for the old Francis Aviation Halloween parties. Some of my students and instructors organized a little nighttime requisitioning party. We built the office inside my hangar without a peep from the authorities. Everybody would congregate there, it was so neat and cozy. With a little heater we could even keep warm in winter. This office worked out marvelously well, but it was quite small.

Soon I had enough business going that dealerships were asking me to sell their airplanes. They wanted me using their planes with all the students I was turning out. I learned that I liked selling. A Mooney distributor asked me to take on a dealership. I went out to their factory to test fly their single-place Mite. The Mooney Mite was nice to fly, their cheapest airplane by far, and inexpensive to operate. However, they had placed the gear control level right along the fuselage. Every time I used it, I lost skin on my knuckles. That was not an airplane I wanted to sell. I took on a Piper dealership instead.

By then I really was doing pretty well. A contract I had with AC Spark Plug brought in some fairly heavy flying. We also flew other freight for General Motors and did a lot of charter work with passengers. I would say the business was fifty-one percent students. The rest was all the other things put together in a changeable mix, including aerial photography, primarily done for the newspaper.

Real estate people would have us take pictures of properties that were going to be developed, and again after they were developed. We also had a contract to photograph every school in the Flint school system.

In the fifties I was flying with my students when Flint was in the path of one of the worst tornadoes in U.S. history. The sky was pink, green and gorgeous, then became black. I rushed all the planes into the hangar and was on my way home when the sirens began to scream. That twister caused some terrible devastation and killed a lot of people.

President Eisenhower even came in his Air Force One, called *Columbine*, to Flint to see the extent of the destruction. While he was there, his plane and mine were the only ones allowed to fly over the devastated area. Every aerial picture of the horrible scene was taken from my plane. I wound up flying until after midnight, helping the media file their reports. Before I went to bed I flew most of the exposed film of the disaster to Detroit.

Flying for aerial photography left some unique moments in my mind. The challenge was a little like flying the old planes for a parachute jump. One time I almost lost a photographer. I didn't realize he had loosened his seat belt and was leaning out the open doorway quite so far. It got a little bumpy. As I saw his body moving out the door, I tipped the plane to the other side and flipped him back inside. We never even talked about it. I didn't want to lose that contract!

Propellers and Parabaloids

When I won a Piper dealership, I had airplanes that couldn't be stored on their noses because they were low wing and tricycle-geared. We kept getting more and bigger airplanes. Eventually I had Piper, Cessna, Beech and Aerocommander. I outgrew my facility, so I went to an architect recommended by Bob Steadman. I didn't want the building to be too expensive, but one that had some kind of appeal for aviation. Tom Sedgewick came up with a hyperbolic paraboloid shape for the roof. I thought, "This is it! It follows the line of a propeller going through one revolution!"

We put that one thousand twenty-four square foot building out for bids. The return bids had a ten thousand dollar difference, from the highest to the lowest. The bid I took wasn't the lowest, but was much lower

Under construction.

than the highest. The group that won the bid was smart enough to build a model of it, so they knew they could do it. Most builders were so afraid of the design, they jacked up the price. The men we hired started building in the middle of the winter. They didn't seem to care if it was snowing or raining, day or night. They were there, and the business of building was done and it was great!

My new building had an orange roof with side walls stained brown to create enough contrast for people to notice coming down the highway. My office was paneled in charter oak and, because I was there so many hours, I had them put in a tiled shower. Outside, it was all landscaped. To me it was the Taj Mahal.

We held a big open house. I sent an invitation to Mr. Piper, Sr. and to my astonishment, he came. By this time, everyone was talking about the space race. I got a female manikin, put a space helmet and a bikini on her and set her in the shower. Mr. Piper thought this was the most marvelous thing in the world. He greeted everybody and took them into my shower. He had planned to go home that evening, but stayed overnight and stayed for the next day, too. I asked him how many people he had working for him. He replied, "Half of them." He was a grand old man, full of fun.

Pipers and Flying Photographers

That party really launched me into the Piper business. They developed a new idea. A business that wanted to sell their twin-engine airplanes had to own one and commit to selling three. I returned to the bank to plead my case: "Do you think you can sell...?" "Oh, yes, I can have them sold in a year."

I don't know where I got all that courage. I put in my order for three twin Comanches, and I soon had them sold. I sold an Aztec fully loaded and an Aerocommander fully loaded, too. I made enough money with them to pay for the planes, pay off the office building and had no debt.

I also finally had an office with enough space for a flight simulator. In studying the attrition of my students, I found that when bad weather caused missed appointments they often didn't come back. So I went to Rudy Frasca, who was building flight simulators. I became very impressed, not only with the simulator he designed, but also with him. I was using Cherokees for training by that time, and I wanted a simulator cockpit that responded exactly like the planes my students really flew:

Proud owner: Trimble Aviation.

with the same stalling speed, cruising speeds, power settings. When Frasca finished my simulator, I had as much money invested in it as if I had purchased another Cherokee, because it was all a special-build.

And, it worked! Now my students would come weather or not. If the weather was good, we'd fly in the airplane. If it was not good, we'd fly in the simulator. Later, when I was on the President's aviation committee, I asked the FAA to accept a certain percentage of simulator time as part of the total required hours for a license. I added a lot of flight simulator time to my logbook with that beautiful machine before I ultimately sold my Frasca simulator to the college in Flint.

Then I sold my business, because we moved to Ann Arbor, where I went to work for Bob Twining as a consultant. Bob Twining was the best boss I ever had. He was successful with charter and sales of late model executive twin aircraft. He never treated me as though I was any different from any of the other pilots at Twining Aviation. I flew pretty sophisticated airplanes for him. To this day, he has a special place in my heart. I haven't known anyone before or after who said immediately, "Here's the keys, go fly." The only plane he had that I did not care for is known as a "push-me pull-you" like the animal in the movie *Doctor Dolittle*. The Cessna 337 is too noisy for enjoyment.

Not to Repeat!

I was not a perfect pilot. There is much that I never want anybody to repeat. One day I took a man and his wife up to fly. We used a Stinson 108 with two places in front and one in back, which flew like a wet log. I was cruising around nicely when the husband requested me to do something special, "Give her a little thrill." I pulled up into a careful stall and that Stinson suddenly fell back into a hammerhead stall, nose down and under. Trying to recover, I quickly moved the wheel forward, then back, but it felt like it wasn't even attached to the airplane anymore! We went into an inverted spin, and there was nothing in my experience to help. (Well, if we are going to crash, we are going to go in hard!) I shoved the wheel forward and held it, then gradually pulled back and, by golly, I had something.

When we came out of that stall, we were close to the leaves, over the turkey farm that had sent us a notice: "Please don't practice over our turkey farm!" There were turkeys flying in every direction. (They'll get the number of this plane, and I'll have to pay for hundreds of birds without eating even one of them!) I looked at the lady sitting in the back. She looked at me, and her neck turned red like the fluid rising in a thermometer. I never said anything. Back at the airport, I made a very soft landing. I never allowed a passenger to dictate to me again.

The Stinson 108 was licensed as a three-place airplane. I learned the hard way that weight in the rear seat pushed the center of gravity out of the envelope. It should have been placarded to avoid stalls when the rear seat was occupied.

Teaching to Learn And Making the Sale

If anyone thinks test flying is any more hazardous, thrilling or satisfying than flight instruction, let them think again. Although the opportunity for military test flights was denied women during my active years of flying, discussions with test pilots compared to my own experience as a flight instructor has lead me to the opinion that they run a fairly parallel course. Both add up to hours and hours of boredom, punctuated by moments of pure terror. My own work involved some testing of aircraft and aircraft components. Although I was not testing the extreme edge of the flight envelope, I was placing the aircraft in unusual circumstances. Those tests were never as much of a challenge as keeping ahead of my students.

When I first started teaching, my boss, Jerry Francis was keen on making a buck but wasn't much interested in maintaining his aircraft. At one time, after four forced landings in his planes, I was so tired of it, I went in to tell him I needed a vacation. "What do you mean you need a vacation? Every day is a vacation for you." He gave me a couple of days off.

I flew seven days a week, twelve to fifteen hours a day. I would get so tired I could hardly stand up. But it was a marvelous kind of fatigue. For a change in routine, I planned cross-country trips with my students on weekends when the flight department was slow. These were more complex trips than the ones I flew as a student. We scheduled them to places where we could have some fun. I would take a student needing dual cross-country time, and students already signed off for solo cross-country flying would fly the rest of my airplanes. This helped them earn the

Miss B surrounded by students.

required ten solo cross-country hours for their private license. We did plan the trips to arrive back before dark, but sometimes it wasn't much before.

One flight as a group was to Gaylord, Michigan, for a canoe trip down the AuSable River. The canoe livery picked us up at the airport to transport us, the canoes, a picnic lunch and sweaters for a four-hour run down-river. Riding the current made us so cold the fire felt good when we beached our canoes at the halfway point. Over the fire we made hot coffee the Finnish way by heating water in an empty Crisco can, pouring in the coffee grounds, and letting the water boil again until the grounds settled. Just before we poured, we threw in a raw egg to collect the grounds. Then we served the best coffee in the world. After everyone ate, we had time for a little exercise; swimming, fishing or just walking down the river collecting stones. Our departure time came too soon, even with the excitement of the return flight ahead of us.

My memories of teaching center on specific students. When I start thinking of these individuals, only the events of their most interesting days spring into my mind. Unfortunately, it was so exciting to be flying on a daily basis, spending time with so many interesting pilots, that I made little record of all that happened. I just went from one full day to another.

A Typical Day

My life as an instructor in the mid-forties through the mid-fifties would start no later than eight, when the first student would wander into the office, probably still eating breakfast. In the days of the GI bill it was occasionally obvious they had very little sleep and were sometimes hung over. I also flew with many regular, paying customers. They intermingled into a happy group of guys and gals, all bitten by the flying bug. Civilian flight students were usually from families with money, while the GIs came with veteran's credit for time served, which amounted to four dollars for each day they served. The amount of time these men usually had earned allowed a full private pilot course, and for some, a commercial license. Some even had enough time built up to pay for the instrument rating. The idea was to train returning vets for entry into the commercial air fleet. This worked in some cases, but usually only for men who already had flight time in transport aircraft.

Returning Vets

One day a young, good-looking lad came into the flight office, plopped himself up on the counter and announced he wanted to learn to fly in a hurry. He told me all about his service in the Marines and all the places he had been as a man of the world without even taking a breath. After telling him to get off the counter and to get both feet firmly planted on the floor, I talked about what our program was and what we expected of him. He reminded me of when I needed to convince Jean Ramsey that I was serious on my own first day of flying. Was he as scared as I was then? I was concerned that he might turn out to be a problem student, but he was no more or less a problem than the rest. Marine Vets seemed very self-confident, ready to take anything, and I found this young man was every bit the Marine, full of chutzpah. Soon after solo, he asked to fly our plane to Arkansas to visit his family. It was hard to convince him that this was premature. I found out much later that, soon after I began instructing him, he was hopping passengers out of a farm field, having the time of his life. It's a darn good thing I didn't know that at the time. It is also a darn good thing he was a good pilot, because he sure was an inexperienced one.

Not all returning vets were as capricious or as ego-driven as this Marine. Bob Vaillancourt, the friend who bought the T'crafts for me to use, was a very special student. When he returned to perform aircraft maintenance at the Buick Hanger, it was his strong work ethic as well as his ability as a mechanic that inspired others and helped him climb the ladder to air transport boss. Getting him prepared to fly for GM was a rare treat. He helped me get started in business, I was helping him fulfill a dream he left behind when they downsized the Army Air Corps.

There were other levelheaded Vets who padded their GI bill income with the job of line boy. Our fleet was larger after the war, but our line boys still had the responsibility of getting the right gas, oil and tires for our training fleet. No matter the size and weight of the planes, they still had to pull and shove them into and out of the hanger, sometimes several yards away from parking. In addition, they still had to hand-start the planes. I still taught students to keep their heels on the brakes, watch for instructions from the 'boy' and do exactly what was called for. "Brakes and contact" meant turning on the magneto switch and holding the brakes, so the boy could swing the prop to start the engine. "And make sure he is out of the way before you take your heels off the brakes!" The line boy job is very important to a fixed-base operation. He is the person everyone goes to with almost every task connected to the fleet. Younger, or older, these men and women exchange their steady hand and enthusiasm for what seems like very little money or for time at the controls of a plane in the air.

FLYING NUNS, DOCTORS AND TWINS

Gerty Prochazka and I taught a nun to fly. She wanted to learn how to fly because the children she taught were so interested in airplanes, she wanted to be able to speak intelligently about them. She looked out of place in her full, long habit with a cowl that included a band across her forehead. Her cape would flap as she walked across the field and the men would laugh. "How are you going to work with that?" My part of the job was to teach her to recover from a spin. This required a parachute. As I cautiously began to explain what this contraption required, she grabbed the 'chute, strapped it on and away we went. I remember another time when she was up, we were having a problem and she just plain didn't want to come down. She was having a grand time with the challenge of this flight.

Some of my students came to me through unusual lines of contact. I remember two different occasions of minor surgery requiring local anesthesia where I talked to the doctors while they worked on me. In both cases, they learned to fly with me, and I eventually sold them airplanes.

It seemed that the advent of the new light twins like the Cessna 310 and the Piper Apache made pilots think that this was the way to go, with more range and safety and more seats for family members. But the smaller engines in the twins meant the pilots had to know emergency procedures very well. In addition, twin load limits could be easily compromised.

My business used an early Piper Apache for multi-engine training. It only had standard fuel tanks. If the plane was not overloaded, it had pretty good performance on one engine. The major limitation was that, on one engine with full tanks, the best the plane could do was a controlled glide-to-landing. That Apache convinced many students that having two engines was not in itself a safety feature and it made them stop taking safety measures lightly.

A gentleman from Alabama arrived at my office one day to talk flying. We fell into a discussion of his need for transportation in his new position as an executive for an encyclopedia company. He said he wanted to fly a twin-engine airplane and preferred to learn in the type of plane he would later fly. Our insurance would not cover his flight training if he started in our twin. As we continued to talk, we figured out what to do. I convinced him to take some initial time in a single engine plane and purchase a new Piper twin Comanche. When the Piper plant called that his plane was ready to be picked up, he was ready to finish his lessons in his own twin. One thing I insisted on was that he never ask me when he could solo that twin. It would be entirely my decision and I would stand for no pressure to rush it. In due time he soloed the twin, with the majority of his student hours in the twin. I lost track of him, but I trust he is still happily flying. He was a fine pilot by the time he left me, and that twin fit him as though it was built around him.

A Cadillac Air

Fairly early in my years of instruction a couple from Northville, near Detroit, dropped by my flight school. The gentleman was a successful manufacturer. He and his wife both wanted to learn to fly and were willing to travel all the way to Flint to take lessons from me. We spent a lot of time together. I even got to know both of their children. Later the family wanted a more sophisticated airplane than one of our training planes. I suggested a Cessna 170B and offered to customize it for them.

They agreed, so I set out to pick fabric, leathers and a paint scheme that I thought would make them feel they had a showpiece. I chose Cadillac fabric and leathers because that was what he was driving. Then I learned that one doesn't just decide to imitate Cadillac without their permission, so permission I sought. It took a couple of trips to GM in Detroit before I had their approval. Then I learned why Cadillacs cost so much. The deep brown tapestry insert came at thirty-five dollars a square yard. The metallic tan leather for the trim was equally outrageous. I had the interior done by the man Piper Aircraft chose to do their show model of the Aztec. It was finished off with my paint specifications: butterscotch, white and Hershey-chocolate brown. The resulting airplane was like none other in the world. The family was tremendously pleased. I can't remember being more pleased with any other of the many planes I refurbished. For years after they sold it for bigger and more sophisticated aircraft, their first plane was still a crowd stopper. They talked most lovingly about seeing it all over the country.

The pilot matures. Circa 1957.

Paid in Cash

Some people say that farmers are a hard sell. I found that generalization is just not true. I remember one farmer who came into the flight office to ask if I had any planes to sell. I said we sure did and took him out to the hangar to see what we then had for sale. As we walked back to my office, he said it was the first time anyone ever took him seriously enough to show him a plane. I asked if one of ours had taken his fancy. "Yes." As I drew up the papers for the sale, I asked how he was going to pay for the plane. He put his hands in his coverall pocket, drew out a wad and paid for the plane even before he knew if he could fly. As we taught him to fly, he put a nice grass airstrip on his farm with a hangar for the plane. I suspect he surprised a few folks.

Another unusual sale was to a pilot from Bay City who came with his wife to look at our inventory. After deciding they wanted a nice Cessna 172, he left, saying he would return in a minute. From his car, he brought back a cigar box full of dollar bills he had been saving for that certain time when he found the right airplane. It took us almost an hour to count the money because it was in small bills, with coins to boot. Still, it was fun to have such an enthusiastic customer. I've seen them on a few occasions since. We always laugh at the memory of that cigar box and its contents.

Not all my students bought airplanes and surely not all were wealthy, but the one thing they all had in common was a love for flying.

Racing: Different Races, Different Faces

Mary Clark and I raced several times together, becoming especially close friends. Mary was so great she was made of twenty-four karat gold. She worked for the Red Cross in the Pacific, was a devout Catholic and was family oriented. An executive, she worked in Jackson, Michigan, at her family's business, the Crowley Boiler Works. Mary flew a Cherokee 235, was voted best-dressed woman of Jackson numerous times, and was not self-centered at all. On her first race with me she got so excited from listening to all the stories of other co-pilots that she woke up in the middle of one muggy night and washed my underwear. I had no change of clothes, so I crossed the finish line the next day wearing wet underwear! Another time when we raced out west, she went out to check something about the plane, broke a small bone in her foot, and never said anything until after the race was over. When Mary and I raced together, we didn't have earphones or an intercom. It was so noisy in the plane that she created three signs to help me understand exactly what she was saying: "Yes", "No" and "Hold your course".

All the women had to race stock models. We couldn't do anything to soup up the airplanes, or at least we didn't do anything that we couldn't get away with. Everyone polished her plane so much, the rain wouldn't even stick on it. That constant polishing became a sort of chuckly thing. We used to say we took the co-pilot along to wash the clothes, do the navigating and keep the belly clean.

It is strange how people behave under pressure. Taking somebody on a race as a co-pilot more than once puts the friendship at risk. Some of the girls racing together just once didn't wind up as friends. It's close contact under pressure and is hard work. I always had a co-pilot I enjoyed. Everybody who flew a race with me wound up being a very good friend. As you have already read, Joan Hrubec and I went on to share many different adventures as pilots and otherwise. Others were part of my life for shorter, but significant, periods.

Lucille Quamby

My co-pilot in AWTAR 1960, Lucille Quamby, arranged for a well equipped Cessna 172 which was advanced for its day, with dual Omni receivers to capture the VOR (navigation) signals. The owner of the C-172 put up our expenses. Lucille was a tall woman, a phys-ed instructor from Detroit, who I met through Ninety-Nines. She was a do-er! We started the 1960 race in California and ended in Wilmington, Delaware. On the way out we met our sponsor in Las Vegas. I became concerned with how much money Lucille was gambling, but she knew her limit and kept to it. I learned during this race that I had to plan ahead for a new time-consumer. Flying a sponsor around before the start used up more time than I imagined. But it was time well spent.

When we were in Las Vegas the sponsor asked us to make a presentation of a Brunswick bowling ball specially made for Betty Hutton. He arranged for us to meet her before her one-woman show one evening. Betty was wonderfully relaxed and animated. She also was so pleased and interested in the air race that she invited us to attend her show that evening as special guests. She placed us in the front row and, when the show began, she sat down on the edge of the stage right in front of us and sang her first number directly to us. Her song was "I'm Just a Girl Who Can't Say No." At the end of the song she had us stand up to introduce us as racing pilots. I was in awe of her. I believe I could have closed my hands around her waist; so tiny, yet she was full of energy. After the show she invited us backstage where we spent a lovely hour with her. Betty was from Detroit and seemed genuinely pleased to be talking with folks from back home. I still believe she enjoyed the conversation as much as we did. That evening gave Lucille and me a lift that lasted all through that race.

Since Lucille had arranged for the plane we were using, her name was listed as pilot-in-command. We arrived too late to correct the record, but I was definitely acting PIC. After we went in to the hotel to relax at the end of the race, we were advised that the weather was deteriorating and we should return to the airport to get our planes hangared to avoid

Lucille Quamby, Betty Hutton, Miss B, and the ball.

damage. Dressing quickly, we felt lucky to find room in an executive hangar. Then the next morning we were greeted with the news that a gas truck had backed into the nose of the plane, smashing the front end from the prop all the way back to the firewall. This plane was the owner's pride and joy. The damage made things rather tense for a while. We couldn't fly the race plane home until the damage was repaired. Finally, someone loaned us an old Cessna Cardinal to get home. It had a geared engine. As pilot, I could not figure out how to adjust the engine, so we barely made it over the hills flying home.

Pat Arnold

In the '64 Powderpuff Derby from Fresno, California, to Atlantic City, New Jersey, my employee Pat Arnold beat me. I should have fired her for that but I couldn't because I sold her the plane she was using for the race. She took fifth place in a Piper 180 flying one hundred thirty-three knots with an average ground speed of one hundred fifty knots. Mary Clark and I also flew in a P-180, but with an airspeed of one hundred twenty knots and an average ground speed of one hundred thirty knots.

The plane she bought from me was almost magical. It was an example of the unique airplane that comes along from the factory by chance once in a great while. For some combination of reasons, one plane in a type can perform unusually well. Pat couldn't understand what made my Comanche so fast. First it started out with all its parts well connected. Then it sustained just enough dimples in its skin during a hailstorm to improve, not destroy, the airflow. I teased her, "Ever see a golf ball without dimples?" With its now perfectly marred skin, this Comanche could almost keep up with the Beech Bonanzas. I was in the business of selling planes so, after getting her hooked, I sold my Comanche to her. Then I watched her win.

When Pat beat me I *really* wished for the money to stockpile a hanger full of fast planes. However, all my race planes were ones I cleaned up after students' use or were from the line that I had for sale, or were stock airplanes on loan for the race. As I compare more than one of the same model that I flew in a race, it was a combination of things that would make one plane faster than another. Sometimes an engine would be just as sweet as anything and other times one was kind of a dog. Once in a while, I'd have an airframe itself that was fast. There are a multitude of things that can add up to more or less air resistance on an airplane. Even the way the wings hang can make a tremendous amount of difference in drag. To slip through the air in a sweet, fast plane is a dream.

I only experienced the kind of magic we found in that Comanche one other time, in a fast P-260 I sold to Sam Stewart, a retired Air Force engineer. When he came back home to Flint, he continued flight training with us. I knew the P-260 was fast when I sold him the plane, so I asked to use it in a race.

I like to think my placement in the races was all due to pilotage, but I really know better! To balance things, some planes, ones that I expected to do well, were dogs for no reason anybody could determine. Note that change? It was the doggy planes that lost races; the pilot just won them.

COMPETITORS

There were a lot of keen women competitors in those air races. They forced me to concentrate to do my best job. I always made sure the airplane was in good shape. The electronics were always the best available that I could afford. If I needed to, I would switch electronics just for a race. I used to practice the take-off to see how many seconds would pass from the minute I hit the throttle to the time the plane was straight and level on course. If it took over a minute, well, I needed more practice. When I was getting ready for a race, I did things that I wouldn't teach others to do. The gear came up maybe a little bit sooner than I would normally be able to risk teaching someone and I turned maybe a little bit sooner. I definitely carried quite a bit more airspeed. Everything that was down and dirty, came up and clean, rapidly! I also always took off as close to being on course as possible. Only if we had a lot of wind would I restrict myself to taking off into the wind.

With all my planning, it still wasn't just the weather that could throw a race. An early Cessna180 had a prop governor go. That shook our airplane so badly it spilled the gyros off. Another time in a Bonanza the air was so turbulent the governor went off again and the prop was running away. That meant I had to throttle back or it would disengage the engine from the airframe. We had electronic problems during some races. When engineers first specified the transistorized radio, they didn't realize they were hotter than the old ones and didn't design for enough air circulation to get through the cowling, so the radios heated up. Sometimes they burned out completely. When the insulation on that wiring burned, it smelled awful!

So many things could go wrong. Each time we started out everything had been checked over most carefully by a mechanic, but racing is altitude flying, it's hot air flying, and quite a bit of the time when we were out west we were in turbulence and dust, which didn't help either. High-speed landings didn't help the tires much. Every single part of the airplane took quite a bit of abuse. The mechanic from my business, Howard Geyer, loved the excitement of the races. He went over every last nut and bolt, every wire, every inch, of any plane I raced. Occasionally he came out to a race. Because I was doing so well, he had girls trying to charm him away from me. You can be sure the man loved that!

Each plane was also inspected before a race by members of the race committee. During a race, if we had any kind of problem that required a mechanic, we would have to contact an official after we landed for the night. She had to be on hand to make sure our mechanic didn't do anything that would alter the stock performance of the airplane.

We tended to fly much higher or lower during a race than we would have flown on a regular commercial run. We carried oxygen if it was to our advantage to fly high to catch favorable upper winds. Depending upon the landscape we were flying over, we were going to be two thousand feet above the highest peak. In the west that can mean flying high where the air is thinner. West to east, we always carried oxygen: north to south, we might not want that extra weight.

The challenge to me when I was flying a smaller, lighter, slower airplane was that faster airplanes in the same race had a little more leeway in choosing when to fly. In a Bonanza or a Comanche pilots could stop and gain an advantage. Sometimes they would not fly at all one day if the tailwinds were not right. They could make it from coast to coast in two and a half days. My smaller planes needed three plus days, but still, with the handicap system, I placed with the slower planes almost every time I raced in one.

All the competitors would generally stay at the same motel at the beginning of a race. Then we were on our own, choosing which designated airport to stop at for the night. Each small decision could make a difference in how much time we spent in the air and, thus, what our placement might be in the final standings. Once we were on the ground, anything that would interrupt the airflow had to be cleaned off. With so many hours in the air, it was a messy job to get the black, sticky, oily film off the wings and the underbelly before we could go to sleep every single night.

When we'd go to the flightline in the morning, there would be dew on the plane that had to come off. The windshield had to be spick and span and the prop polished. Suddenly the announcer would say, "Ladies, man your airplanes!" and we'd always run... for the bathroom, always with a line ahead of us... ."Hurry, hurry!"

I had a feeling of total elation once the gear was up and we were on course. "We're on our way!" Each leg stop was determined, first, by the range of our particular airplane. Notice that it was the limit of the individual airplane that determined the length of a flight, not the limits of the pilots. The only thing I remember doing to myself to prepare for race day was to cut down on liquids. From four o'clock in the afternoon of the day before a race, I cut my fluids way back so I would be able to miss potty stops. We didn't use any ventilation during a race because open vents cause drag. We sucked on lifesavers because our mouths would get so dry. It always got so hot in the plane, we became dehydrated. We wore the same clothes day after day, landing twice, or sometimes only once, during the day. We'd saw the handle off the toothbrush to save weight. I felt we also got faster as the race went on and I lost my usual ten pounds.

When it was to our advantage to stop during the day, or for the night, we'd look for other pilots. There was always a place designated for us to stay at authorized airports. By rigorously enforced rules, we ended each race day, for sure, at sunset. Somehow then a new mood took over: the camaraderie of everybody pulling in at the stops, everybody wanting to know how everybody was doing and nobody telling the truth. We shared the biggest cock and bull stories! "God, I just came through that weather and there was nothing to it." The next pilot would say, "It was just terrible, we had to go a long way out of our way." "We were going very fast" And the next, "We may not even make our handicap." You couldn't believe a thing anybody said. We always ate together, chewing the rag for a bit, then everybody went off to bed because the morrow was another early rising with more of the famous green eggs and ham.

Co-Pilots, Not Passengers

Every time we landed at a designated airport, the co-pilot had to race to get our time log stamped. She ran to get the stamp, climbed back into the plane and made sure she had her door closed and locked before the plane lifted off the ground again. It was a frantic few moments, dangerous especially for the co-pilots.

Mary Clark racing for the time stamp.

Time on the ground didn't count against us as long as we completed the race during the allotted race days. Still, sometimes you would pull into a stop barely long enough to get gas. If we found a stretch during the time of day when the winds were the highest and from the most favorable direction, we didn't want to leave that envelope of good air. If we absolutely had to come down, it was only to get back up into that air as quickly as possible.

Most of the time my co-pilots were handling the book work and the E6B. This is a mechanical computer, a series of circular slide rules, where we line up part of a formula on a series of circles and get the answer to the problem on another part of the dial. I'm glad students are still taught how to use the E6B in the best training programs, even with all the modern equipment. Using the E6B makes them understand the formulas. If I needed a frequency to cross check for our location, my co-pilot would give me that frequency. Again, students today need to learn how to do a crosscheck and keep up on all the old skills that make flying accurate and safe. A pilot never knows when or what equipment will break down, whether just flying around the patch, or racing.

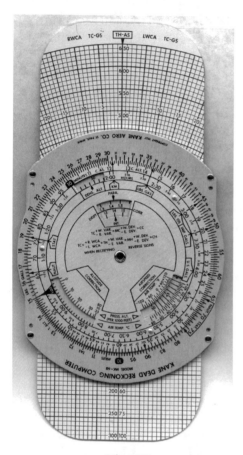

The E6B.

Before getting back in the air every race morning, I laid out the true course. I would select a heading to achieve my true course based on the recorded and expected wind direction. Then I recorded the distance of each leg and the time it should take to fly each leg for the entire day. I marked the race chart in fifty-mile segments, then made smaller marks every ten miles so my co-pilot could make frequent ground speed checks by timing our flight over landmarks.

With the constant work of speed checks from the time I leveled off, my co-pilots really were busy. A big percentage of my navigating was by dead reckoning because our best racing was in a straight line. At the altitudes we usually flew, distance-measuring equipment (DME) wasn't useful. We flew parts of many races so low, we were in ground-effect, where no kind of navigation but dead reckoning would work. Even that can get ticklish if you're over an unfamiliar area. Flying low, the terrain seems to go past all the faster. One year we won averaging 212.9 mph. When a plane is only five hundred feet above the ground (AGL), that's landscape, twhew, twhew, twhew, going by. Even flying the mountains, we might fly low through the saddlebacks. Then we really watched the sides of the mountains as well as the ground. In racing, the wind determines what the real shortest distance is between point A and point B. The pilot who wants to win takes all the risks she can tolerate.

My co-pilots and I were all flat-landers. We got into the mountains and it was always a strange experience. I finally took advantage of a demonstration of downdraft and other challenges to mountain flying when a director of aeronautics asked if we would like to see some mountain sheep. After he scared the sheep off that mountain, I understood when and how to let go, when to sense a recovery point and when to put on the power. Finally, during one race, we heard mountain pilots talking about how to fly over flat land. Visibility that day was recorded as ninety miles. Ninety miles! When the visibility is that clear, your sense of distance changes. From the air, we could see so much territory it was hard to judge the relationships between landmarks.

We were often surprised by what the press might say about us. Sometimes we were so tired we might have said something that contributed to a strange twist of words in the paper. Once I read, "A woman who called herself a Flying Farmer from Flint won again." Well, I was never a farmer. I picked strawberries for my grandfather, and we called our place in Montrose a ranch, but we only raised trees. I have quite a collection of clippings, including the amusing misunderstandings and downright fabrications and misquotes from the races, mostly stacked in folders. The best-organized clippings and other memorabilia are in scrapbooks created by Mary Clark. Her mother was another woman who spent time on her knees, praying. She would go to mass for us all the time we were on our way out to a race, as we flew the race, and on the way back from a race.

All co-pilots had to be able to think as fast as the PIC did. At the end of one race we experienced a mechanical malfunction that might have caused us to land gear up. "Just save the trophies!" Instead, my co-pilot calmly offered additional suggestions for what we might try, and we kissed the earth with our blessed wheels. As I said, my co-pilots were made of good stuff!

We still used to tease them something fierce, puffing ourselves up as the pilot-in-command. However, we knew that in reality these women were invaluable to us. They not only did all that work, they had to be qualified to take over without a second's notice if something tragic were to occur to the high and mighty PIC. Yes, I treasured my co-pilot, my assistant, my insurance, my SIC. And, I made sure they were just as capable of being high and mighty as I was.

Hang On Janey Hart

(with Janey Hart)

I met Janey at a Ninety-Nines' meeting when she was very pregnant and couldn't fly in the races. She said if I would use her plane, her Padre Island Resort would sponsor me. I flew in one Powder Puff and an International race using her F-series Bonanza. This model was from the first year Beech went into the heavy skin without putting in a more powerful transition engine. As a result, the entire F-series of planes were not fast. We tried all sorts of things to make that plane faster, even re-hanging the wings. I placed the plane in the top ten, but never better.

Sometimes I even forget that we did win once using that plane. In a Flint Journal article from 1957, Edna Gardner White is shown giving us a trophy with the caption "Bernice Trimble Forms Own Flying Service". Joan Hrubec and I had flown Janey's Bonanza in the one-hundred-seventy mile Michigan SMALL Race from Lansing to Traverse City. This was an efficiency race. I won, flying one-hundred-twenty-seven miles an hour in the darn thing. With the later series Bonanzas' reputation for speed, we were so slow they called the company to see if I had somehow cheated during the race.

Flying Reynold Wrap

Most of the planes I raced were mine, cleaned up right off the flightline, but the only other Bonanza I raced in for an AWTAR was again an early model owned by Doctor Zeis. It had such a lightweight skin, we'd look out in turbulence and watch the skin wave at us. His plane was not for the faint of heart at high speeds in rough air. I pre-planned each race as near to my intended flight path as I could, but that year with Janey, nothing went the way it should.

In rough air, the flex in the light skin of Doc's "Reynolds Wrap"

Bonanza popped the door open. In addition to the loss of speed caused by that drag, a Bonanza isn't supposed to fly with the door even ajar because it is an integral part of the strength of the airframe. We all learned how ugly it is to race from east to west against the wind during the 1959 Transcontinental but for Janey, that flight from Lawrence, Massachusetts, to Spokane, Washington, seemed endless.

(Janey Hart) The door would not remain closed! It would pop open just enough to ruin the aerodynamics of the plane. This was a race, after all, so I had to hold it shut ...all the way across the country. Every time we landed somewhere, my fingers would have to be pried off the door handle.

According to the Ninety-Nines' history book, 1959 was my fourth AWTAR. I'm listed with 7,000 hours flying time, an airline transport license, flight instructor, and single and multiengine land ratings. Janey was flying in her third AWTAR, had over 780 hours, and was "one of the few women with a helicopter rating."

The 1959 race was the only AWTAR I flew against the prevailing wind, from east to west. It was exciting to finally command a plane in a race that had the potential to skip a fuel stop or two. But that year the door, the direction of the wind and our range all conspired against us. I would file for a distance I expected to make...and have to stop short. Our airplane was a problem, the length of the race was a problem, the course was a problem and the weather was a BIG problem, especially up in Montana!

Our first clue to what was ahead came as we got to altitude when the weather began to shroud the tops of the eastern mountains. Then, heading for Helena, Montana, we flew through a pass where we watched some Bonanzas land on a rancher's airstrip. We could tell that landing there would take us out of the race. "Oh no, that's not for us!" As soon as I spoke, the weather became worse. As I turned toward a light spot in the clouds, we ran into some hail. I mean it hailed hard, nearly knocking us out of the sky. Before we could react, we rounded a mountain, flew into sunshine and Janey wondered, "what the good Lord is doing about everybody else?"

We later learned a number of the racers had landed in that pass. A young gal ran the airstrip where we watched the Bonanza's land. She had not been married very long to her rancher. Out there in the middle of Montana she didn't have other women around, so she really entertained the gals and put them up for the night. Some of the planes had to be pulled onto the roadway for takeoff once the weather cleared, but they finally all managed to get out. Those pilots had such an interesting time that many of the "mountain women" correspond with their benefactor to this day.

I think I raced four times. My first race was in the old Bonanza with Babe Ruth, then I got the twin and raced with B. Seeing a business advantage, B was serious about racing and took the time to groom her competitive edge. Most of the time when we competed against each other, I didn't do very well. I was pleased then just to be able to race. One time Nan Rudolf raced with me, before we were required to have a licensed co-pilot. That was a fifth time I competed. All of my racing was in the All Woman Transcontinentals, never an International.

I stopped being a contestant when I was asked to be on the Ninety-Nine's

Race Board. I could have continued to race, but by then, I had an Aerocommander, which is hardly a racing aircraft. I set up the clocks, took them along to the next stop, did some other stuff, then packed the clocks up when it was all over. I took my son Walter along to the races when I was flying the Aerocommander.

The Stoneage Viscount

Back in the days of Janey's first twin, a Bonanza, named the "Stoneage Viscount," she called me one day to ask if I would take her plane and fly Rose Kennedy around Michigan. She would be my co-pilot. Mrs. Kennedy was campaigning for her son, Jack, for President. She was a very dynamic person. She would get into the plane, put blinders on and fall asleep until we arrived at the next stop. Then she would go like mad, come back, fall asleep, and be ready to go again. We pretty much made a complete tour of Michigan before she was done.

Mrs. Kennedy never regarded me simply as the pilot of the airplane; I was part of her entourage. It was most enjoyable. When she would speak, I would go into the back of the room and listen to people, then tell her what people were saying while she was busy working the room. She thought my sleuthing was great.

I flew a lot of candidates during my career, including Thomas Dewey and Michigan's Governor Swainson. I even helped the Republicans campaign when I flew beside Eisenhower's airship, the *Columbine*. The *Columbine* is a beautiful Constellation. It is now a museum showpiece.

The Stoneage Viscount! Yes, that was a big old thing. If duct tape had been available then, I would have put it all around the windows. Those windows made a dreadful noise as they whistled and screamed.

When I decided I wanted a twin-engine airplane, my good friend figured out a way we could fly almost every twin in the state for free. B had dealers fly them in from all over so we could compare them. After we recorded lots of extra hours in the logbooks, I decided on the Stoneage Viscount, despite the noise. It was a fine choice for me at the time. I could get all eight of my kids in, plus a lot of freight, and thunder off in whatever direction I wanted. We went to Mackinaw Island in the summer. I'd drop the boys off for camp on Walloon Lake and fly on to St. Ignace. We had a house up on the East Bluff in those days. I could fly to Washington with the children all in the plane, to spend time with Phil. One time I filed IFR although there wasn't a cloud in the sky. By the time we were over West Virginia there was so much smog we would have had to file anyway. Now sometimes I wonder about taking my kids down to breathe in that Washington soup.

I also became a chopper pilot because, although he was Lt. Governor when he first wanted to run for the US Senate, Phil was not widely known. There were only three helicopters in Michigan at the time. A road contractor owned one. His pilot was my instructor. I paid the insurance and whenever the contractor didn't need a charter, I could use the chopper. I'd pick it up from Detroit, get it to Lansing and begin to work the campaign schedule by collecting Phil, flying to rural areas to attend fairs, then flying home.

One of the fundraising prizes at the Michigan State Fair was a ride in the chopper with me. An older man wished he hadn't won. I told him I wouldn't do anything scary. He never said a word the whole flight. He didn't say thank you after the flight either.

My license was #25 Whirly Girl. I flew a few other choppers while I was in Washington, a Huey and a Sikorsky. Whirly Girl, what a ridiculous name.

Beyond the Stoneage

Most people know Janey best as a wife, mother and activist, I asked her to share some more flying adventures, so people will understand why I would recommend her as an astronaut candidate.

After the Aerocommander, I had a twin Comanche. I also flew with Phil in that. He was a wonderful co-pilot. He'd sleep the whole way, or else he'd wake up and ask, "What town is that?" "Well I don't know, because we're not going there anyway."

One time we flew down to the Virgin Islands. We had to stop in West Palm Beach to get our clearance paperwork done, because we had to get fuel in Great Inagua in the Bahamas on the way to the Virgins in those days. When I called on 121.5 that we were coming in for fuel, Great Inagua replied, "We haven't got any." I said I'd go to South Cacaos. "South Cacaos doesn't have any either, because the same ship delivers to us both, and it has a leak or something."

Fortunately, a Bahamas Airliner was listening. He thought he knew of a man over on Turks Island who might have some barrels of eighty octane fuel in his house or garage or storeroom... or something. He suggested that I land there. I was glad for this advice because I only had about forty-five minutes of fuel remaining. Turks was a missile tracking station run by PanAm.

When we called they said yes, they had the fuel, but it had to be saved for an emergency.

"Well, this is an emergency. If I can't get in there, I haven't any fuel to go anywhere." "Can you wait another ten minutes? We're about to fire off a weather rocket." "That will be fine, but then I'll be right in."

When we landed, we found we were only the twentieth tourists on the island in that entire year. We spent the night at some man's house and were treated to a party at the governor's house. Turks was under British control then, staffed by poor, early-stage foreign service personnel and a middle-aged doctor and his wife who must have done something dreadful, because they were isolated there. All these people were actually quite lovely to visit with, so we had a wonderful time.

The next morning a truck came to the plane with barrels of fuel and, thank God, I had a cloth to strain the fuel through. He only knows how long those barrels had been sitting there on that tropical island. As we finished straining the fuel into my tanks, the man who controlled the fuel arrived with his assistant and stated they were going with us, just like that. I loaded everyone into the plane and flew on to Puerto Rico. The last I saw those two men they said, "We are off to Miami."

My circumstances gave me opportunities then as well as challenges. I had a wonderful lady working for me from the time Walter was born in 1950. She helped me provide a stable environment for our family. She worked for me until Phil died and the children were gone. Now she lives in a nice retirement apartment. I couldn't have done any of it without her. We call her even now about the cooking. We were all good friends. We respected her; she respected me.

Circling to Land

Janey is one heck of a good pilot. I remember when she had an Aerocommander adventure above Lake Erie. The 500-A was not a tiny thing, but mid-size, multiengine. Again, she proved that it is best to keep calm in crisis, fly the plane and act quickly in a logical sequence.

I was coming to Michigan from Washington, D.C. having picked up my niece in North Philadelphia. It was IFR until we got to the edge of Lake Erie. I noticed one engine kick a couple of times, so I made a little note, figuring it was the fuel injector. When I got into clear air, everything about flying cleaned up nicely, naturally. Later I seemed to be using an excessive amount of fuel. I computed the remainder and had enough for the flight. Later still, when I computed fuel, it was normal flow.

At about four thousand feet over the city of Pontiac, Michigan, we were plenty high enough when one engine quit, then the other. No engines. Too much silence. I quickly looked at everything on the panel. I especially looked at the fuel gauge right away, but it was recording eighty-five gallons. I'll never forget that number.

The control tower had been built in Pontiac by then, but it was not yet manned. My niece was a little alarmed. "We are okay, Suzie, the airport is right there."

But, my God, the traffic! I decided I was going to fly straight in instead of getting into the pattern with the other planes. As I circled down to pattern height, I called on 121.5 that I was coming in with no engines. They said somebody was in the way, so instead of going with the traffic, I had to circle around to fly in against the traffic. Landing with, instead of against, the wind uses more runway. This emergency left me with just enough runway to bleed off the airspeed to settle the plane down. "Fly the airplane, fly the airplane, even onto the ground." We were thankful to stop just before we hit the grass.

Maintenance came out with a tractor to haul us in. My gauges still said I had plenty of fuel. The men in Pontiac checked everything under the cowl and couldn't find anything wrong. So we looked farther out from the power source.

At the time of this problem, after refueling the Aerocommander, a lineman would replace the cap, turn the handle, and a spring would load the gasket in place. Back in North Philadelphia when the guy filled the tanks, he apparently didn't notice our spring had snapped. Once I was flying, there was enough airflow over the wing to start sucking the fuel out on the negative-forces side of the wing. With rubber tanks, the bottom of the tank was sucked up to give a reading of gas that wasn't there. In Pontiac, when they took the gas cap off, the bottom of that tank popped right out of the opening!

Fortunately, B and I had already planned to leave the next day to take my brand new Aerocommander back out to Oklahoma for a run through their famous familiarization course. We told them about my experience and suggested that they ought to re-engineer their gas cap.

They agreed. This Aerocommander was filled from a single port, using an o-ring for the seal. The fuel bladder had been attached to the bottom of each wing. When there was no seal, the gas bled off the top of her high wing so it couldn't be seen. Eventually the suction pulled the tank off the tie down and forced the fuel out of both wing bladders.

With the circling Janey had to do over Pontiac, she was so close to the ground witnesses were concerned she might clip a wing on her turn to final. It was quite a piece of flying.

Politics

Because he was in politics, occasionally I would ask Phil ahead of time about something I planned to do, not wanting to embarrass him. He would say, "Well certainly, go right ahead." When I went off to Vietnam or something Phil would laugh, "I wonder what your sisters are thinking." They were very conservative.

The trip to Vietnam could have embarrassed Phil a lot in the Senate. When I got back, I asked if there had been a problem. He admitted they'd given him a little chitchat in the dining room, asking why he couldn't control his wife. He would laugh like hell. One day I finally said to the dining room at large, "All right, which one of you wouldn't have gone if you'd had the chance?"

He supported my decision to visit such a controversial place, even as the mother of his eight children. He also supported me when I went to Albuquerque for the astronaut physical.

Mercury Women

When the Mercury Astronauts were introduced to the world in 1959, Castro defeated Batista and President Eisenhower imposed mandatory quotas on oil imports as distances between countries were shrinking. Vice-president Nixon opened an American exhibit in Moscow and Queen Elizabeth dedicated the St Lawrence Seaway. The first U.S. nuclear-powered merchant ship was launched, the first turbo-prop airplane went into service and American Motors introduced the compact Rambler to compete with the German VW Beetle; both were stick-shift only. *The Miracle Worker*, *Raisin in the Sun* and *Anatomy of a Murder* were challenging the established viewpoint. Pantyhose and Barbie dolls entered the American scene.

Flint, Michigan

Back home, the normal thing for a girl to do after high school was still to get married, and start a family. Guys would work for GM and gals would work for Bell Telephone Company or one of the banks, until their children started to arrive. Then most of them retired as soon as they began "to show", if not before. Very few women did other things straight from school unless they were like Berniece, tops in her class, with a job in a law office. Dedicated girls attended years of nursing school for the honor of taking orders from doctors and earning little pay. If the daughter of a well-to-do family went to college, it often would be as much to catch an educated-man-on-the-fast-track-to-security as to get an education. Locked in their dorms by 11:00 pm, giggling under the watchful eye of the housemother, they used to plan how to earn an "Mrs." degree and start a family. The average female homemaker still could not get a loan on her own.

I was an established businesswoman in Flint, a commercial pilot and an air racer. I was also newly married. I already knew Jacqueline Cochran's reputation as the premiere woman pilot in this country and as a businesswoman. At the Cleveland Air Races she loaded perfume into a plane and sprayed the bleachers with it. She also used to give each of the contestants in the AWTAR some of her special cosmetics, all set up in a tube a little bigger around than a pen. It reminded us to present ourselves as ladies and promoted her business. We took the part of being a lady seriously. There was no question that a person could be fired then for "conduct and decorum" beyond the norms, and the norms were stricter for women than for men. Airplane doors were high off the ground, but when we stepped out of our airplane at the end of a big race, we were in skirts, with our hair arranged and our lipstick refreshed. Most often, my co-pilot and I wore matching outfits. I had just begun to hear about Dr. Randolph Lovelace and his clinic.

LOVELACE AND COCHRAN

In *Ladybirds*, Henry Holden and Captain Lori Griffith use a quote from a February 1960 *Look* magazine: "Air Force Brigadier General Don Flickinger said without elaboration, 'Women would not be given serious consideration for space travel until three-person space vehicles were in use.'" I didn't know this, but I agree that, "Flickinger, however was telling a half truth. [Because] he and Dr. W. Randolph Lovelace II had already decided to screen some women." And we were serious about entering space!

Dr. Lovelace, who eventually asked me to call him by the nickname "Randy", was known as the developer of the pressure-release for parachutes. He also helped develop a pressure suit so pilots wouldn't black out at high g-forces and studied the effect of high altitudes on humans. In the late '50s, he formed a group to create the criteria for the selection of the first astronauts, the men who became known as the Mercury Seven. The tests they developed were to check first if the body was normal, then to see how it reacted to certain stresses. Jerrie Cobb was the first woman to take the tests, in 1959.

Holden and Griffith continue, "In August 1960 Lovelace let the secret [of testing women] out at the International Space Symposium in Stockholm, Sweden. *Time* Magazine quickly followed with a story about Cobb's progress." In the paper delivered at the Symposium, Lovelace also stated the reasons the doctors wanted to test women: "Women have lower body mass, need significantly less oxygen and less food, and may be able to go up in lighter capsules, or exist longer than men on the same supplies. Since women's reproductive organs are internal they should be able to tolerate higher radiation levels." I never knew why they wanted to test me, I just wanted to become the first woman to enter space.

Then the press had a field day. "Bachelor Girl Cobb 36-27-34." Can you imagine being judged for space by such measurements? How dumb do they think people are? Soon, Flickinger was quoted again. "The tests were preliminary, Cobb isn't an official astronaut." The Brigadier's judgement was made on the basis that, "NASA didn't have any space suits to accommodate their particular bodily needs and functions." Well, how did NASA come up with suits that met the bodily functions of men?

This set the stage, but still was not all of the truth. In 1961 Jacqueline Cochran (Jackie), was the most advanced woman pilot in our country. Among her two hundred and fifty speed, altitude and distance records, she was the first woman to break the sound barrier at Mach 1 and Mach 2. As an established champion of other women in aviation, she financed a top-secret study to include women in the Mercury database. The physical cost $1500 per person. These tests became the first basic, full medical documentation of a group of healthy women. Prior data on women came from clinics when they were ill. Lovelace wanted to plot the results for women on the same graphs as the results from the men. My records show Floyd Odlum (Jackie Cochran's husband) covered these tests with a donation of stock (Letter on file in Eisenhower Library), and later the proposed flights to Pensacola were covered by him with cash. (Noted in margin of letter from Jackie dated September 1, 1961.)

During one of the races, I heard some pilots talking about testing women for astronaut training. Both Jerrie Cobb and I were in the race. The idea of women in space was exciting for us to contemplate. Jerrie was working for Aerocommander then, setting altitude and speed records, becoming very famous. I learned later that Jerrie was the first woman to be tested in the Lovelace Program. She then helped review all the women with higher ratings, pilots who met the physical restrictions. She says she chose to recommend me based on my demonstrated range of experience.

In August, 1960, Lovelace began sending letters asking qualified women pilots if they would be interested in participating in his tests for a woman astronaut in the space program (see Lovelace letter, p. 144). Many women were interested, but I don't know how many applied after the first round of letters.

On November 28, 1960, Jackie wrote, "Dear Randy, Following our discussions, I have been thinking over the plans, still in the formative stage, for 'The Women in Space' program. ...I think the Government will get more benefit out of the early phases of this program, which for the most part will be research or study phases, ...if the entrance requirements for volunteer candidates are liberalized.

"Out of the 20 women pilots whom you contacted, 7 did not meet the written requirements laid down by your preliminary medical examination. Some of the remainder will not pass the physical tests and still others, for one reason or another, will drop out. The consequence is likely to be that the group carried on into succeeding phases will be too small to reach adequate conclusions as to women as 'astronauts', per se or compared with men. ...

"At this stage at least, ...you should bear in mind, ...that it is very likely to be a long time before any one or more of the candidates is put into space flight."

Jackie goes on to discuss married vs. unmarried women, to encourage testing of younger and older women (beyond the thirty-five years old limit), and attitudes and responses she expected between different groups. She refers to the WASP program of WWII, and says: "...the best material was between 20 and 23 [years old]."

In closing, she worries about jealousy and criticism. "Everyone in your original group that is passed medically should be certified at the same time." In a supplementary note, she urges 'Randy' to accept the twin sisters from Los Angeles together as valuable to the program, noting that he had tested the twin with the fewest flight hours first.

Under a Clinic letterhead dated January 31, 1961, Lovelace replied with a list of fourteen more women who he would like to test, including me! Five of the women on this list were to pass the tests, including me! In his letter, Lovelace stated he prefered "to have girls who have 1000 or more hours of flying time." My copy of this letter also has an attachment that lists various aviation experiences and the personal details we had to divulge, including such things as lineage, church affiliation, operations, languages, degrees, research interests and military experience. "Including me" at this point clearly demonstrates that the focus of the selection process was mostly on flying ability.

I learned I was selected for the physical testing just before a sailing trip with Janey Hart. Letters and telephone calls flew back and forth between Cochran, Lovelace, and me. I told them I'd recently had hepatitis. I was afraid it would knock me out of the testing program. Lovelace said to come anyway, that first they would make sure that every candidate was basically healthy. My first test would be the Bilirubin test, which would show if the illness had damaged my liver. If I failed that, I would not be tested further. Actually, if any candidate failed any test, they could not continue.

The letter from me to Cochran dated May 6, 1961 (see page 134), refers to Alan Shepard's bounce into and out of space. I was so excited I didn't even spell "tremendous" correctly. Though more experienced, I still was the girl who had not taken typing in high school, and there were no computers to spell check for us in those days.

Twenty-Five Pilots

Many years after the Mercury Program I learned that the twenty-five women who met the initial Mercury criteria arrived at the Lovelace Clinic in Albuquerque, New Mexico, by twos, on progressive Sunday nights. (This total has been debated for years. The number I use here is agreed upon by people who have researched the question.) They began taking the tests Monday morning and headed home late Friday night, unless there was a need to retake any test for clarity of the results on Saturday. This was quite a bit of time to pull out of a competitive business schedule, based only on chance. The women who were selected to take the Lovelace Clinic tests covered the same age range as the men who were tested; height, weight, experience, and demonstrated intelligence. We came from all kinds of social and aviation backgrounds. It is my understanding that most of the women candidates had more flight time than the Mercury men, in some cases by quite a bit, although none of us had jet time, because women weren't allowed to fly military or commercial jets in the fifties. In the fifties women were not so rare in the sky anymore, but the men still seemed to feel a jet wouldn't stay in the air if it wasn't flown by the right gender.

Tom Wolfe captured the excitement and the effort of the Mercury Astronaut program in his book *The Right Stuff* in 1983. The excitement

· SPECIALIZING IN PILOT TRAINING ·

BISHOP AIRPORT
PHONE CE 2-9766
FLINT, MICHIGAN

May 6, 1961

Miss Jacqueline Cochran
Cochran-Odlum Ranch
Indio, California

Dear Miss Cochran:

It gives me great personal pleasure to have been selected by you and the Lovelace Foundation for the Research Project, "Women In Space". I must admit after passing the physical at the Foundation, I feel pretty healthy.

Thank you so much for all your generosity in assisting this program. I sincerely hope your wishes and hopes for this project will be more than just gratifying.

I am most anxious to hear what the next step will be. It is very difficult to with-hold the information what with to-days tremdous event.

Thanks again.

Respectfully,

Bernice T. Steadman

of the time, however, didn't just involve the men to be sent into space. It seemed to involve the rest of the world in a search for all kinds of the right stuff for the astronauts to use. For example, they used ideas about how to deliver food from the traditions of Denmark to the outback of Australia, and my country seemed to be at the forefront of the search. The reach for the best, most efficient, and safest materials seemed to tie us together in a common goal. This race for space was the same kind of Everyman effort we produced in WWII, when Rosie the Riveter was just about as important as GI Joe. A second similarity to WWII was that pilot-astronauts were flying so high and so fast they acknowledged their dependence on their ground crew as an invaluable part of their team.

I was caught up in the same state of excitement as when my high school senior paper, a requirement for graduation, was written on aviation. The excitement was again shared by schoolchildren when they were challenged to write papers on rocket propellants and to design equipment for space travel. In a science paper, my co-writer designed lightweight footgear based on snowshoes for the deep dust some authorities predicted covered the moon's surface in a uniform, deep layer. Out on the point near her home, another girl from her class shot all sorts of unstable, oddly propelled objects into the South Arm of Lake Charlevoix while safely protected from the blast by an overturned picnic table.

The excitement seemed to grip everybody. TVs were set up in schools so Russian successes could be worried about and/or applauded. At night we watched for falling stars and Sputnik, the first salvo in the race for space. There was a sense of relief when our new president, Kennedy, stated we *would* win the race to the Moon. In the magic of this time, even Republican children were encouraged to believe in their Democratic President. Adults also believed in their president and their country. For proof of adult trust, I offer this condensed list of the invasive tests given to one of the Dietrick twins from L.A. (Although I cannot find my own list, I remember my testing schedule as being the same as this one.):

Sunday, January 15; No restrictions on eating or drinking...but do not smoke after 7:00 p.m. or eat until after completion of blood volume... .
    Collect first stool sample... .'
    Nothing to eat, drink or smoke after midnight, or until... .

Monday; Nothing to eat, drink or smoke on arising.
> 7:30 a.m. Deliver stool specimen to Laboratory Appointment Desk on first floor of Lassetter Laboratory Building.
> 8:30 a.m. Report to Physiology Department, first floor of Clinic Building for blood volume determination.
> 9:30 a.m. ...report to Miss Thomas for referral to Cardiology Department for EKG and Master 2-step. Also...instructions for preparation for proctoscopic examination at 1:30 p.m.
> 10:30 a.m. Report to Audiology Department, Clinic Basement, ...
> 12:00 noon. Lunch.
> 1:30 p.m. Report to Miss Anderson, Clinic Emergency Room, for proctoscopic...[run down the hall with the enema bag].
> 2:00 p.m. Report to X-ray Department...for lumbar and dental x-rays.
> 3:30 p.m. Report to Dr. Seacrest, third floor of Clinic Building.
> Evening, No restrictions on eating, drinking or smoking...
> Collect second stool sample...
> Shampoo hair; do not re-apply any hair dressing until after EEG.
> Nothing to eat, drink or smoke after midnight, or until... .

Tuesday; Nothing to eat, drink or smoke on arising.
> 7:30 a.m. Deliver stool specimen to Laboratory Appointment Desk...
> 8:00 a.m. Report to Dr. Howarth in Radiation Therapy ...
> 11:30 a.m. Report to...for BSP liver function test.
> 12:00 noon. Eat only a light lunch, prior to exercise tests.
> 1:30 p.m. Report to...for exercise tests.
> 3:00 p.m. Report to...for referral for electroencephalograms [EEG].
> 4:00 p.m. Report to Dr. Bivens...
> Upon completion of examination by Dr. Bivens, report to...for instructions for preparation for gastric analysis and colon x-rays...
> Evening. No restrictions on eating, drinking or smoking...
> Nothing to eat, drink or smoke after midnight, or until... .

Wednesday; Nothing to eat, drink or smoke on arising.
> 8:15 a.m. ...gastric analysis [swallow 3(feet of rubber tubing].
> 9:00 a.m. Immediately following.... report ...for colon x-rays.
> 10:30 a.m. Report to physiology Department...for pulmonary function testing [ride the bike].
> 12:00 noon. Lunch.
> 2:00 p.m. Report...sinus x-rays.
> Upon completion...report...for instructions...preparation for stomach, chest and esophagus x-rays on Thursday morning.'

2:30 p.m. Report to Dr. Wood...for eye examination... [four hours—bright light recovery and eighty-three more].
Evening...free except for preparation for stomach x-rays.
Nothing to eat, drink or smoke after midnight, or until... .

Thursday; Nothing to eat, drink or smoke on arising.
8:00 a.m. Report to...for stomach, chest and esophagus x-rays. ...you may eat breakfast. [if time permits]
10:30 a.m. Report to Carco Air Service for flight to Los Alamos. [Total body counter—K_{40} determination.]
2:30 p.m. Report to Miss Thomas for referral for phonocardiogram and vectorcardiograms.
4:00 p.m. Report to Dr Merideth, ENT Department for ear, nose and throat...[frozen inner ear].
Evening. No restrictions on eating or drinking or smoking this evening.

Friday; You may eat breakfast, but do not smoke until after cold pressor...
9:00 a.m. Report to Dr Roth...cold pressor tests [arm in frozen water].
10:00 a.m. Report to Dr Fountain.
12:00 noon. Lunch.
1:30 p.m. Report to...for tilt-table examinations.
3:00 p.m. Report to Dr Seacrest.
Upon completion of Dr Seacrest's examination, report to Miss Thomas to see if there are further instructions or repeat tests."

There is a similar list of the tests the Mercury astronauts took in Jake Spidle's book, *The Lovelace Medical Center: Pioneer in American Health Care*. We are mentioned on page 139 in the book under a mis-dated photograph: "After the successful experience with the project Mercury astronaut candidates, the Lovelace Foundation tested a group of women pilots to establish their suitability for space flight."

We were found to be suitable.

Notice how many times parts of all of the candidates were x-rayed in one week. Would anyone do that today? I learned to be wary when they didn't mention the name of the test I would be taking, and I became a little uneasy whenever I was sent to see Miss Thomas for instructions.

But before I submitted myself to this torture, I took off with Janey Hart for an island-hopping adventure.

The Caribbean Interlude

(with help from notes by Jane Howard)

I was just getting over a relapse of hepatitis when Janey Hart called to invite me to go sailing with her. Her sister, Susan Fisher, and friends, Betty Maloney, Louise Hyde, Jane Howard and I made up the rest of her crew. We were a mixed cast of characters: an elegant New Yorker; a woman who raised thoroughbreds; a fashion consultant; an aviation instructor; and two politicos. Four were experienced sailors; Betty and I were not. We were to spend the month of March, 1961, sailing the Caribbean aboard a leased auxiliary; a gaff-rigged, fifty-six-ton steel ketch called the *Harebell*. I needed a lot of rest, orange juice and sunshine. The invitation was just what the doctor ordered. Bob and I were in the process of moving, but he said it would be of most help for me to get away and return as a healthy woman.

With Janey as pilot and myself as co-pilot, we left Michigan in a hurry, because we would have been stuck behind some nasty weather. We flew into Washington International aboard Jane's *Elegant Bird* just in time to attend a dinner party in Georgetown. I remember our hostess, Mrs. Dixon, seated four or five to a table and directed current event questions to each of us. Herblock, a famous cartoonist, was there. Everybody seemed so self-confident and was associated with some position in the government. It was a stimulating experience, a wonderful start to our adventure

Then we had to wait a while for Louise to complete some business. Once she was ready, the weather was miserable, so we all continued to have fun, bunking at Janey's until we could at least feel safe flying on instruments. Finally, we were all settled into the pale bone-colored, metallic green-trimmed Aerocommander 500A with storks painted on the sides of it. That airplane was beautiful. It was one of the planes I sold Janey over a period of years.

In those days it was still unusual to see a plane full of just women, especially one that had the more complex twin-engines with no man at the controls. As we waited for IFR clearance at the end of the busy runway, the passengers and crew of a neighboring Capitol Viscount saluted us.

Betty, the daughter of Senator James Couzens, was a bit frightened of flying. She shortened the time she would spend in the air by taking a train to Florida. Betty's mother was even more cautious. The only way Betty could calm her about the trip was to say, "Oh, Mother, we won't be flying. We'll simply take the Couzens' Causeway between the islands." Betty beat us to Miami on the train. When we finally arrived, she had our rooms all set so all we had to do was visit together until the wee hours of the morning about what to expect on our trip.

Then we began to fly Betty's causeway: to Grand Cayman; to Montego Bay in Jamaica, where some of us submitted to the smallpox vaccination required to proceed further; back into the plane, on to San Juan; finally we landed on Antigua for our last hot water for two weeks, a trickle shower! From Antigua, "Couzens' Causeway" was to lead us via Martinique to Grenada, where we would meet our boat and crew in St. George's Harbor. However, we had to land on Martinique for some small repairs, which made it too dark to land at Grenada Airport, so we flew on to Trinidad. Then we had a little difficulty arranging an inter-island commercial flight back to Grenada, but when we saw the approach to the airport there, we were glad we'd flown to Trinidad. Landing in the dark between the cliffs beside an airport only lit by torches would have been a shock, and not only for Betty.

Finally, we had the first view of our beautiful, temporary home. Abiking and Rasmussen of Germany had built the *Harebell.* It was owned and operated by a former U-boat Lieutenant who hated to use the marine diesel engine as much as we did. Datelieb Von Iberg had a magnificent beard and an English wife, Joanne. Captain Von Iberg enjoyed his time as a Hitler youth, but when he said he liked boats he was removed from his family and trained for life aboard the U-boats. As the war progressed, he turned his boat over to the British, lived as a prisoner on Joanne's parents' farm, wooed her successfully and took them both off to island life tempered by the soft westerlies of the Caribbean. The *Harebell* also had a cook and three deckhands.

Our short time on Grenada was beautiful. The first sail was back to Trinidad, which looks like a spread-eagle.

As we sailed on from port to port we learned some local history. On an island administered by the Dutch, a man who heard that we were all pilots invited us to see his building project. It was a mystery how he heard about us, but we went over to see his airplane. It was still under construction: the wings were in the bedroom, the fuselage filled both the living and dining rooms. Although the pieces were already fabric covered, we could tell his creation was well built.

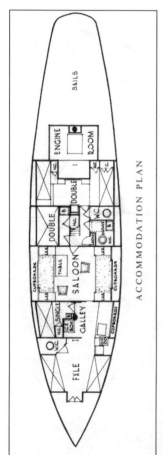

Curacaiou is where a rowboat brought ice and a little boy came alongside to sing a rather shocking, naughty calypso song. We also attended church there in a cathedral on the hill. The local bishop asked, "Did you ladies come to revolutionize our island, or are you just passing through?" Everywhere we went, everybody seemed to know all about the six ladies sailing the Caribbean alone. We couldn't figure out how they knew, since we had no radio. Even school children came to look at us. This trip was before tourists flew in for short visits. The natives were only used to sailors who arrived bronzed by the wind-driven trip from the mainland. Our white skin against the white sand attracted the children and made them reach out to touch us.

Living on board the *Harebell*, we had no shower and the head was so complicated you needed a helicopter rating to work it. Whenever we heard a flush we had to run in and help open one cock and close another or we'd have a flood all over the cabin.

When I picked oysters off some mangrove roots with Louise, we found an old sailboat and a pith helmet. We never learned the story behind them, but such adventures were common in the Caribbean.

The days just slipped by. We passed an island called "Prune." As we sailed on to Bequilla Island for St. Patrick's Day, we caught a barracuda for dinner. The trip to Tobago Keys was beautiful. Jane Howard later did a watercolor of the main island with a ship tipped to its side being cleaned of barnacles. In the background, she painted Princess Margaret's honeymoon cottage. It is still hanging in my house. We boarded a horrible ship called the *Seabeck* for cocktails with some very nice people. The ten-hour sail from St. Vincent to St. Lucia was incredible. We lost our jib, saw a porpoise, and watched a whale swim on the horizon (which was close enough for me!). Just at sundown, we saw a green flash caused by phosphates in the water. At St. Vincent, we taxied around a botanical garden, but the St. Lucia harbor is beautiful just as it is.

Susan's foot became infected. The doctor told her to stay out of the sun and the water. I am allergic to sunlight. I slowly became covered with awful cold sores. Betty's marvelous designer sun hat was now secured with a shoestring. Only Louise remained efficient and beautiful. But then, Louise always seemed to look marvelous. One time she came to visit me in Montrose, asking to see the "Ranch". Our eighty acres were level back to the house, then they sloped down to a meandering creek where we farmed blue spruce. Picture the two of us in my little black MG getting stuck in the mud near a side road. We tried to work our way out. Two men stopped to help. When the truck we hired finally pulled us out, I looked like I'd crawled through the mud. Louise, who worked just as hard as I did, looked elegant, as if she had just stepped out of *Vogue*.

Sailing is a little like flying. I really only remember the unusual things about our trip. Aboard the *Harebell* with reduced responsibilities, a lot of the trip was just a feeling of peacefulness. I remember a story Janey told us about one night on a different trip, when she was sleeping offshore. It was such a calm night, she woke up and went back on top. There was no moon, not a cloud in the sky, no land lights anywhere. The boat wasn't making a ripple. She experienced a complete horizon, encircled by the absolute mirror of a bowl of stars. Without a sound, she felt the people on her boat were the only people left in the whole world. Out there in the center of sea and sky peacefulness, she felt that wasn't a bad thought.

One of the most unusual experiences I remember about this entire trip was Pigeon Island. Like a piece of property in the wilds of Canada, Josset Legh, a British showgirl, held ownership to an island via a ninety-nine year lease. Everything she had was hand built by herself and her daughter: thatched huts, a thatched-roof kitchen with no walls and the main lodge where we stayed. One afternoon while we rested aboard the *Harebell* before sailing to Marigot Bay for a swim, the captain of a Buccaneer ran "Rest in Peace" up in flags. Our captain replied by running up a bra and a pair of rose-patterned panties! Of course, we were invited over for cocktails on the Buccaneer after that!

At Marigot Bay Jane went snorkeling and saw a shark. That discouraged the rest of us. At an evening picnic on the beach, Josset cooked things from the island: conch and all manner of seafood. It was delicious, but the only thing I could recognize was a bit of tomato, or was it pimiento? We watched the moon and stars come out. Finally, as we dug our oars in to row back to the *Harebell*, the phosphors in the water formed sparks.

Pigeon Island was a marvelous experience in itself. Josset was telling me, the one she considered already to be an astronaut, that I needed to tell the authorities that the island was full of strange noises in the dark; Russians or creatures from space. She was right about the noises. The island is honeycombed limestone. When the wind blows over the openings, it creates whistles that can be carried from place to place by the tunnels. One morning I woke early to hear something like singing. I went on deck and watched a group of fishermen chanting as they worked their nets on the beach. Their chanting was carried to me through the honeycomb. We all left Josset's full of good food and fine memories.

On Martinique at Diamond Rock, we saw a school of whales. This island is a bit of old France. The French culture seems to bring an extra softness to the air. We followed a road above cliffs overhanging the sea which were covered with fern and coconut palms. After two weeks of sailing, we finally went ashore for a shower at the marina. First to feel clean, Betty and I took a cab (an adventure in itself) to see a hotel which looked quite magnificent on the side of the mountain. That road cost $68,000 a mile to build and employed every able-bodied man on the island. We got out of the cab, walked into the lobby, spoke to no one, and returned to the harbor, where everyone knew where we had been before we got there. We made no reservations, but when we returned up the mountain to the *Lido* with the group, they had a table set exactly for us! That was a strange, but marvelous, dinner.

One day out from Martinique, Janey and I sailed the *Harebell* while the captain went below for tranquilizers and the rest hung on to their bunks. Then after Dominica and Guadeloupe we all decided to sail through the night: at the wheel one hour, off two. Although the wind was not too much to handle, we sailed on a pitching, changing sea. In the salon, a gimbaled table rode at a forty-five degree angle most of the way. We used bunk-boards and pillows to wedge ourselves in, trying to sleep. My hour was the last before daybreak. When they say it is darkest before the dawn, I can tell you it surely is. With the deck awash, I had to hang on for dear life while I tried to read the compass. (Imagine a novice doing that!) Otherwise, I only had the occasional glimpse of the North Star, Polaris, to steer by. We were trying to hold ten degrees off Polaris. It was a great challenge, a tremendous night I'll never forget. We couldn't land at Antigua until daylight because they only used torches to mark the harbor.

On Friday, when we finally returned to dry land in English Harbor, I called Bob and learned that a letter had arrived from Dr. Lovelace which set the date of my participation in the Mercury tests, beginning the following Monday. I boarded a British West Indies plane to get back to Michigan, settled what I could into our new home, packed a few things and took off for my next adventure, in Albuquerque.

LOVELACE CLINIC
4800 Gibson Boulevard SE
ALBUQUERQUE, NEW MEXICO

March 24, 1961

Mrs. Robert A Steadman
Trimble Aviation
Bishop Airport
Flint, Michigan

Mrs. Steadman:

Examination of potential women aeronauts is continuing. We have reviewed the credentials you have sent in and find that you are acceptable for these examinations. Miss Jacqueline Cochran, who has had extensive experience in high altitude, high speed flight and was in charge of the WASP program in World War II, has been kind enough to review the entire program and has made a most generous donation, through the Odlum-Cochran Foundation, to the Lovelace Foundation that will take care of the expenses for room, food and transportation up to $200 during the approximately six days you would be here. She will serve as a special consultant in this program.

As you know from previous correspondence, there will be no charge for your examinations by the Clinic and Foundation Staff. We would like you to plan to arrive in Albuquerque on Sunday, April 2 and report to the Clinic Monday at 8:00 am, without anything to eat, drink, smoke or chew (i.e. gum) after midnight Sunday. You should be through the examinations by Saturday noon. Please let us know if this fits in with your plans.

There will be no announcements or releases on this program until all of you have had your check-ups and then, with each participant's consent, only the names of those that pass the entire examination will be released. It is hoped to have the candidates that pass the examinations meet together late this spring.

Sincerely yours,

W. R. Lovelace

W. Randolph Lovelace II, M.D.

WRL:sh

We have had to change your April 23 appointment to April 2.

I was invited!

The Real Stuff

(with Janey Hart)

When I arrived in New Mexico, everybody knew who I was. They knew where I was supposed to be, when I was supposed to take my tests and where I was supposed to stay. There were no choices; I just seemed to be along for the ride. I was used to having my flying business regulated by various governmental agencies by this time, but the Mercury tests were the first time I chose to be exposed to a governmental program set on a strict structural basis. The entire process seemed remote and strange.

It wasn't until I arrived at the motel in Albuquerque that I met Jerri Sloan. We were to go through the tests together. Even though we had separate rooms, we called each other Roomie right away. We were the only pilots being tested at the time. Jerri was married and had a son. I was married by then and had a son. We became great friends during the tests. It is a friendship I count on to this day. We talk on the phone often and occasionally, wonderfully, we get together.

Jerri's husband was not happy about the program,

Jerri Sloan

not happy at all. He would call her, and she would go into a slump. I mean it was not nice. She was so enthusiastic about this challenge, it would have been better if he had been supportive. Instead, at the end of the program, Jerri's husband met her at the airport with divorce papers.

Fortunately for both Jerri and me, my husband was enthusiastic. He thought we had a great opportunity. He did a lot of teasing over the phone to help relieve the stress.

Rise and Shine

At Albuquerque we were in the hospital by seven o'clock in the morning. The tests took all day. I took eighty-four different tests in all. The eye examination alone lasted four hours. This was all part of trying to find out just how healthy we healthy women were. Our doctors were still talking about the Mercury men who had gone through just ahead of us. We learned we were to be subjected to the same testing and selection process as was used on the men. Everything we did was graphed with what they had done, so there was a comparison for all of the data.

One day we were to ride a bicycle in time to a metronome while they monitored our heart and blood pressure. At certain intervals they would change the inclination to simulate a fifteen-degree uphill climb. We still had to pedal in time to the beat of the metronome or, if we didn't keep up with the beat, we might have to wait until Saturday to take this test over again. With the trip to the Caribbean I had been gone so long I was anxious to get home. I called my husband the night before the test. "I really don't know if I can do this test. I haven't ridden a bike in so long." He reassured me. "Just remember, when you get really exhausted, you have a second wind." I wasn't too sure I knew what he meant but when I was going uphill and uphill and uphill, getting really tired, I remembered Bob's advice, took a deep breath and remarkably found my second wind. Finally they told me to stop. I know I did better than the Mercury men because there was a lot of talk about my length of time. I was happy not only to pass, but to have a bit of fuss made about my level of achievement.

We were actually experiencing some of the opening application of nuclear medicine at the Lovelace Clinic. In another test we drank radioactive water. Then they flew us to Los Alamos to a very secret, secret place. When we landed our clearance papers were carefully checked. Then we were whisked into a little building, went down four rapid flights by elevator, walked down a long hallway past a lot of animals and, finally, faced a big set of doors. The big doors opened, we went in, and the big doors closed. Inside the room was a great big THING. Before we could grasp what it was, we were sent into a small room to strip, absolutely strip off everything not permanently attached to our bodies, and put on white gowns. Then we learned we were to be tested by a machine called the Human Counter. As the predecessor to the MRI, the one we entered occupied the entire room and required wooden scaffolding just to reach the opening. When we were ready, we walked up the scaffold, stretched out on a piece of metal and were slid into the machine. Just before we went into this THING, we were handed a chicken switch. "If you get claustrophobia in the Human Counter, push this switch and you'll eject." Well, once inside, we couldn't move. I felt like I was a bullet in the barrel of a gun. Electrons shot into our bodies from all angles. Outside, on equipment that looked like it belonged on a submarine in an undersea movie, lights blinked and proceeded to evaluate a number of human

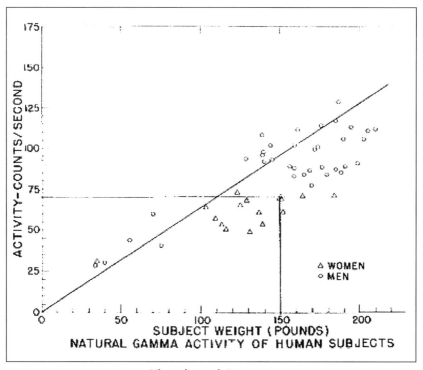

The only graph I ever saw.

body factors. When I left the THING they told me, "For your bone mass you should weigh... so you are overweight by... ." Heck, I could have told them that without the tests!

One day at the end of our examinations, Dr. Lovelace invited Jerri and me to talk with him about the whole program. I took advantage of this chance because I was interested in how we were doing in comparison to the men. I asked to see more of the results of the tests than others did. I felt the results would determine whether women were going to be accepted into the astronaut program. It was during this visit that we learned some of the men did not do as well as we did on parts of the physical. Some of the men also shared a god-like opinion of themselves. Of all the doctors I met out there, Lovelace seemed to be the one most interested in testing us. He and a Dr. Seacrest were most intrigued by our results. Lovelace felt the female body was better suited for space flight and to face long periods of isolation than the male, psychologically as well as physiologically. He wanted to prove the point. If his ideas were correct, he was sure NASA would accept women, our group of women, into the astronaut program. He had no control over NASA's selection, but he felt they would accept his recommendations. Honoring Cochran's request for silence, Jerri and I quietly went home.

Jerri's son David was so enthusiastic about what she did that he went to school after she passed the tests and told everyone his mother was going to be an astronaut. They gave him a bad time about it. So Jerri got permission to break the silence, went to his school and told them about the program.

After I finished the tests, I suggested that my Caribbean buddy, Jane Briggs-Hart, should go through them. Janey had given birth to nine children and was in tremendous physical condition. She seemed to me to be an ideal person for the tests. She was also the oldest of all the women tested. They were about to close the program when I recommended her. Janey and Gene Nora Jessen were the last people tested during the Mercury Program.

When Janey went to Albuquerque, she must have talked more about what the tests were. Their answers also must have meant more to her than they do to me. She understood exactly what a test was for before she took it. In comparison, I just asked what to do and blindly did it. The memory of drinking that eight ounces of radioactive water now makes me shudder.

Janey Goes to Albuquerque

(Janey Hart) When Bernice left us in the Caribbean to get tested, I decided I'd better get tested. Nobody else in the group had anywhere near as many children. I was the geriatric member of the women's group, but I was the same age as John Glenn. The testing we experienced was exactly as it was presented in Tom Wolfe's book, The Right Stuff, and more. We women also had a gynecological exam.

I don't think they do any of those tests anymore. They know that if you are physically fit enough to fly an airplane or drive a car, you can go into space. But this was before they knew what to expect. Some of those tests made me wonder what in heck they expected us to do in space.

I met Gene Nora Jessen in the motel across the street from the clinic in Albuquerque just before my tests. That motel looked like a no-tell motel. Gene Nora and I were very close friends by the time the tests were over. It was good to share experiences and debrief with someone who was meeting the same challenges.

We revealed a great deal more than we intended, but NASA didn't know what kinds of things would have to be tolerated by the human body. So the men and women just went through everything the doctors could think of to test our physical and mental reactions to stress.

I went through thirty-two tests at least. Some of them were just uncomfortable. and some were very strenuous. I think the one where they put us on the bicycle with an EKG hook-up, a respirator to test what we exhaled and an ECG wired up to us all at the same time was the most strain. We pumped with a metronome to keep the beat. About the time I thought I just couldn't make it anymore, they put a sandbag behind the wheel, which made a stress like going uphill, then set the timer another fifteen minutes. When we were exhaling much more than we were inhaling it meant we had achieved total exhaustion. Then they helped us off. And everything we exhaled went into a bag to be measured for whatever kind of gas each individual was exhaling.

The bike test had something to do with the heart and lungs and their capacity under the pressure of prolonged exertion. Again, they didn't know what we were going to be exposed to in space. The heart was simply another muscle to be tested.

I never saw the results of that test. The only results I remember seeing were of the lean-body mass test, done at Los Alamos, where we went all the way into the atomic thing that looked like an iron lung to measure how efficient each body was with respect to the total mass.

Graphing the data was an innovation of the Lovelace Clinic. They established norms when they sent the men through. We were graphed on each of the tests to compare our results with the Mercury men because they were the only standard that existed at the time. We women pilots did very well measured against the men. I don't remember every test, but when they put the electrodes on and gave the shock... !

Yes, they made our fingers jump. I never understood the purpose of it. I know it's the ulnar nerve they were testing, but I also know it had a much more profound meaning to the doctors than just checking our ulnar nerve.

I think they were checking the time from stimulus to response. I remember the test where they had an oscilloscope set up with a camera. The lab had already stuck me with these number ten...

They were thick as fire matches!

The needles were as thick as number ten nails. I was all hooked up, but the camera didn't have film in it. Nobody knew how to put the film in the camera. I told them to leave the needles alone, bring the camera over and I'd load it. I didn't want them pulling those needles out and pushing them in again! I loaded the camera with one hand dangling to my side. It was the most uncomfortable of the tests...well, almost."

I thought the most uncomfortable one was lying on the bed, with a blood pressure cuff on one arm and your other hand in a bucket of water...

With ice. Thirty-two degrees Fahrenheit.

Thirty-two exactly. My hand was in water with floating ice for fifteen minutes. This shoots the blood pressure up because the whole cardiovascular system contracts immediately. After the fifteen minutes we'd take our hand out and they would measure how long it would take our blood pressure to return to normal. "Okay, now how long will it take the other side?" The first side was bad enough, but by then I knew how bad it was going to feel, so the second side was much worse.

I felt like the marrow of my bones was frozen, which it probably was. They did nothing to help our body warm back up. Nature had to do it. They wanted to see how long it would take. They had to be darn sure our cardiovascular systems were working before we took that test.

OUT OF GRAVITY

They did the vertigo test, too. They'd see how long it would take us to get vertigo and how long it would take to recover. We lay in something like a dentist's chair and turned our head to the side so they could shoot thirty-two degree water into our ear canal to freeze our middle ear, watching our eyes to see how long it took them to begin, then stop, pivoting. It also measured if we were prone to seasickness, or any kind of airsickness. Vertigo precipitates this. I wouldn't want to be bouncing around in space with a bad case of... ."

Most of the astronauts have encountered some form of motion sickness. They don't know exactly why. According to the tests it shouldn't be happening.

It may have to do with being out of gravity. While humans are drifting around, their organs aren't being held in place like they normally are on earth. One doctor doing surgery on the brain contacted Stan Mohler after these tests because Stan published a paper on motion sickness based on what he thought the Clinic discovered from this test. He used data from all the men and the women. The neurosurgeon's questions dealt with circulation in a particular part of the brain. When he combined the data from the tests with the experiences of the men in space, the conclusion was that astronaut motion sickness probably involves the inner ear.

WERNER VON BRAUN

Within a year or two after Janey Hart and I had completed the astronaut physicals, I read that Dr. Wernher von Braun was expected in Flint. Janey flew into town so we could try to wrangle invitations to hear his speech. Von Braun's place in history was secure as the designer of the first ballistic missile and the leader of the team that placed the first American satellite into orbit (Explorer I). He also is given credit as the designer of the Saturn I, the Saturn IB and the powerful moon rocket. Unfortunately, the large attendance of GM executives sponsoring the event was traditionally male only. Even the mayor of Flint was unable to get us in to hear Dr. von Braun speak.

The day he was to arrive Janey and I were sitting in my office feeling unhappy about being restricted again by our gender, when I got a call from my line boy. "A man by the name of Van Brown called to see if we could put his plane in the hanger overnight. While I was trying to find out what kind of plane Mr Brown had, the Control Tower called to say Wernher von Braun would be arriving in ten minutes! We raced to move planes out of the way as he taxied to my door. Although it was thrilling simply to meet him, we would not be able to store his Aerocommander in my hanger because it had such a high tail. This didn't seem to bother his pilot, however, so they tied his plane down on my section of the ramp.

Janey and I took the opportunity to introduce ourselves and welcome Dr. von Braun to Flint, telling him of our interest in space and about the physicals we had taken at the Lovelace Clinic. We also mentioned how sorry we were that we would miss his speech, and offered to take him to the auditorium. As we went into town, our passenger seemed tremendously interested in what we had done in Albuquerque and appeared to be so pleased to be riding with us that we were having a great time. Suddenly I realized I was almost out of gas, with no stations near my route, so I sweated the rest of the trip, but we made it all the way to the front door.

As he got out of my car, Dr. von Braun told us that if we weren't allowed into the meeting, he wouldn't go himself. We became concerned about creating a scene and told him of the stag nature of his hosts. He repeated that we were coming with him. He would not go into the auditorium until he was sure we would be included in his audience. He ended the discussion by asking his pilot to escort us to the back of the room near the projector. His pilot was to stay with us to make sure we were allowed to remain. If that didn't work, Dr. von Braun would demand that we sit at the head table with him. I hastened to assure him that seats beside the projector would be great and finally accepted the fact that if we didn't want to disrupt anything, we'd better follow him. He walked through a group of men glowering by the entrance to the auditorium who obviously did not share his enthusiasm for our presence. We settled near the projector and became enthralled as he spoke about the progress of his Saturn V, the rocket that was to take our astronauts to the Moon.

After the speech Dr. von Braun extended a gracious invitation to us to visit Huntsville with the promise to personally show Janey and me around the space facility. I have always regretted that I did not make an opportunity to visit Huntsville before his death in 1977. The warmth and friendship he displayed on that evening in Flint made a wonderful impression. Janey and I were very fortunate to have the chance to meet this great and humble man.

It's interesting how NASA and the whole space program have been realigned. Way back in the sixties Dr. Von Braun felt we didn't need to risk men in space. We could do everything we needed at the time with a telescope placed outside the atmosphere. As far as sending people to the moon, well, he knew there were many things we still needed answers for here on Earth. As long as some of the challenges in space could be more easily solved by other means, he felt we should use our money to understand Earth first.

I'm not sure I agree. Machines are fine for parts of space exploration, but some things cannot be done by machine. We must know the effect of living in space, to be ready to lift off when the time comes. People have to remember, when we were being tested to go into space, we did not have any idea of what our bodies and brains would have to do. It was a completely foreign experience. Would our bodies turn into mush without gravity? Now they know it is the bones that are the problem and they are experimenting to reduce loss of bone density.

Miss B and The Experimental Aircraft Association

There is another von Braun story in my memory banks: I've met a number of people who belong to the EAA, but Herb Dean plays more then a walk on role in my life. When we met in 1961, he told me about a delta-winged airplane he was building in the basement of his house. I was very much intrigued by the sound of this project. When my husband and I went out to see it, we found one of the most unique airplanes of my experience. It was an all metal, two-place, delta-wing pusher with a retractable landing gear; a thing of beauty. He was proud of his plane and of the engineering design work he had accomplished to build it. He had earned his qualification as a private pilot just so he could fly his own design.

Herb was the ultimate EAA member. Proud, independent, a perfectionist and hard working. It was hard not to believe in him. I learned he had even built his home based on building this plane in the basement, with the understanding that he would have to remove one wall to get the plane out. There seemed to be no end to his resourcefulness, but I still worried about the advanced design and his lack of hours. His superb workmanship did not remove my serious reservations about what the flight characteristics would be for such a delta wing. He arranged for a static wind tunnel test.

I talked long and hard with him about the kind of flight experience he would need. I also took him on two or three charter flights with the Aerocommander to familiarize him with not being able to see the wings and to help him evaluate his abilities as a pilot. I even proposed flying the first test flight for him, but he would hear none of that. The dart was his baby and he was certain he could fly it when it was ready.

Herb's delta was in my hangar by the time Dr. von Braun came to Flint. Noticing the attractive little plane, he was willing to do a walk-around. During our walk, the doctor observed that the plane's propeller was too close to its control surfaces. This could adversely affect control of the plane. I passed this information on to Herb, but he still did not seem worried.

As his first flight in the plane came closer, Herb and I searched out as much information as possible on flight characteristics for a delta. Everything we learned seemed to indicate his design was flyable but would likely be somewhat unstable. I continued to present everything I could think of to convince him that the relatively few hours he had as a private pilot did not give him enough experience to attempt the first flights, but Herb had been through so much in designing and building his plane, he was steadfast in his resolve. When the FAA gave clearance to the plane, Herb was ready to fly. His test plan called for a regimen of taxiing to insure ground handling followed by a simple liftoff, with a planned level flight around the pattern before his first landing. He talked with a pilot with some delta-wing jet experience who advised him to hold the plane on the runway for a longer period to insure reaching takeoff speed. I would fly chase plane in my Piper Comanche.

On the day of the test, my husband, Bob, and Mrs. Dean were in the back seat of the Piper and a reporter for the *Flint Journal* was in front beside me. We came around the pattern flying slowly to approach his position on the east end of the runway. When I radioed him that we were coming into position, Herb commenced his takeoff. The results are hard for me to record even today.

When Herb applied full throttle, his little plane moved straight and steady down the runway like a rocket. It left me feeling like we were standing still and it caught me by surprise. I believe Herb was holding the controls forward as suggested by the delta-jet pilot, and it is obvious to me now that he went through takeoff speed so rapidly that he went into negative lift. The result was that, as his speed increased well beyond the seventy miles an hour he needed for takeoff, negative forces held his plane to the runway. I don't know what speed he reached on the ground, but the Comanche was moving well over a hundred miles an hour and he was leaving us behind. When the delta didn't lift off I heard Herb say, "no soap," as he must have throttled back. Suddenly the little plane popped up right in front of us, then pitched forward and stalled. It looked like the Mig-29 that does the cobra aerobatic maneuver during airshows. Herb caught this first stall, rose off the ground and came up almost straight at us. As I swerved to the left to avoid a collision, his delta nosed over again, hitting the ground and killing its creator. We were a sad group in the Piper.

Later, when Herb's delta was rebuilt, I called the new owner to warn him to arrange for a dynamic wind tunnel test. This determined that the interaction of the propeller-delta wing configuration on Herb's plane would tear its wings off. It is now a museum piece.

Speaking of experimental aircraft, sometime after we met Von Braun, I met one of the pilots from the X-craft Program. I feel that if the x-flights had continued instead of NASA spending time on a capsule, the space program could have moved directly into a controlled-flight vehicle. When the decision was made to go with the capsule, it ended the x-program. A person would have manned the x-craft that were being designed then. Our first manned space program would have been an actual flight, and the astronaut would have landed instead of being dropped into the ocean. No trained pilots like the thought of only being a passenger.

NASA went back to the x-program to develop the shuttle. In my opinion, if they had listened to the pilots in the very beginning, they would have saved billions of dollars.

Cochran and Loveless didn't give us the names of the women who failed the tests. That kept the pressure off us. We were told right at the outset that it was a secret program and we weren't to discuss it with anyone. Jerrie Cobb was known because she had a big spread in *Life* magazine. The Dietrich twins were also involved in some early publicity. Six months after the physical was over, *Life* magazine also had a spread on all of the Mercury Women. Then we learned that thirteen of us had passed the tests and who we were. We also learned we had graphed remarkably parallel to the Mercury men. We were superior in handling isolation and, in measured stress, there didn't seem to be any difference between the men and the women. When women are included now, they are basically selected on the medical data we provided. Every piece fits into the puzzle.

I have relied on that astronaut physical twice now myself. Once I had a hip problem. X-rays showed a tumor in a femur and they were ready to put me in University Hospital for a new joint. "Wait a minute! There's no part of me that hasn't been photographed. Go back to the Albuquerque stuff." When they followed my request, sure enough, the tumor showed up there, several years before. Later, when I developed a brain hemorrhage, my doctor had an additional interest in me because I was the first person he'd known with any past, measured neurological data, pre-trauma.

Tethered Mecury

(with Janey Hart)

Janey Hart, Jerri Sloan and I were done with the Mercury tests in Albuquerque by July 1961. Once Alan Shepard blasted off in the capsule, the desire to speak about what we were doing in the program intensified to an almost overwhelming degree. Jerri and I were on the phone constantly, comparing notes, trying to make our program get going by pushing waves of words over the land lines.

Cochran and Lovelace

Meanwhile Jackie and Randy were firing off letters to each other and there was some leakage to the press about our program. A letter from Floyd Odlum shows how he supported both our program and the activities of his wife. He writes, "Saturday, before departing for Paris, Jackie asked me to send you [Lovelace] the letters she had received from the following [eight women]. Jackie said that of all the letters she has received as a result of the *PARADE* article, these are the only ones worthy of your attention."

A letter dated June 16, 1961, from Jackie to Randy lists some of the plans for the Mercury Women and begins to show growing concern for the program:

...those who had their physicals earlier...should...have some "bring up to date" medical checks... .

How long will the group be at Pensacola? ... .

Jackie goes on to write, "I assume that the tests at Pensacola will be for research purposes. But I also assume that if these tests lead to further stages...the girls who report to Pensacola should be committed to the further stages... . The further stages may be for a year or two or possibly even more." Then she closes, "It is apparent that one of the girls has an 'in' and expects to lead the pack. ... Favoritism would make the project smell to high heaven. ...This should be a serious quiet project from now on. [When was this not true?] ...Also it would be a mistake for a participant to be on a special status with NASA or any other branch of the Armed Service."

I really don't want to point out who Jackie may have been referring to. I also cannot be sure that this is what led to a strained relationship between Jackie and Jerrie Cobb, but in a *Life* magazine article, published in 1968, I noted: Jerrie was "...appointed [after Shepard's flight] as a never consulted consultant to NASA Administrator Jack Webb." So this may be the favoritism that Jackie was concerned about. But Jerrie was the first one of us to pass the tests, so some of the singling out seemed natural to the rest of us.

In a letter to Jackie dated the next day, Randy refers to correspondence about the *American Girl Magazine*, replying that "information is not available as the tests are not complete." Then he mentions two letters "with reference to Pensacola" listing nine women, including me. In the second of the letters, we were told that Jackie would pay our transportation funds if we needed them. Our housing in bachelor officer's quarters would be two dollars a night and meals in their mess cost less than a dollar. Today I can hardly believe how little that trip would have cost us. Dr Lovelace also suggested that we plan to get together after the tests in Pensacola to decide on publicity. I was looking forward to measuring my flying skills against other pilots who passed the physical. Instead we were not to meet any more participants until the International Women's Air and Space Museum (IWASM) held a reunion for us twenty-five years later.

A second letter from Jackie was sent to nine of the thirteen women who passed the tests in Albuquerque. Note that we were to have two more weeks of tests and that, "There is no astronaut program for women as yet." She also adds, "It is possible that my previous experience with group efforts by women in the air can be of some value or help in connection with the possible 'Space Program for Women'." Well, she had my attention and my vote! And I was thrilled that I would be meeting Jackie again in Pensacola. We were also warned again to avoid publicity.

Lovelace followed this with an August list of things to study by September seventeenth: "...the FAA manual, mathematics, theory of

July 12, 1961

Miss Bernice Steadman
Trimble Aviation
Bishop Airport
Flint, Michigan

Dear Miss Steadman:

Dr. Randolph Lovelace, II, of the Lovelace Foundation for Medical Research and Education, has notified you of the invitation to go to Pensacola, Florida for a two weeks series of tests to start on September 18, 1961. I strongly urge you to go.

The necessary expense that you will have will be for transportation to and from Pensacola and for meals you will have during this two weeks period, estimated at $2.00 per day. You will live in the equivalent of officers quarters and meals will be provided at cost.

I have agreed to donate to the Lovelace Foundation the needed sum to cover the transportation and meal cost for all of the candidates if and to the extent they need such assistance. If you need and want money for these purposes please, therefore, write to Dr. Lovelace at Albuquerque and such funds will be provided, in the same way that I handled the costs of the medical checks of a majority of the candidates at the Lovelace Foundation.

Approximately twleve women have been invited to take these further tests at the Navy's Aerospace Medical Center at Pensacola. As you probably know I am not a participant in these medical checks and tests. They were set up for women under forty years of age. Some of you may therefore wonder why my great interest and my assistance.

There is no astronaut program for women as yet. The medical checks at Albuquerque and the further tests to be made at Pensacola are purely experimental and in the nature of research, fostered by some of the doctors and their associates interested in aerospace medicine. No program for women has been officially adopted as yet by any of the governmental agencies. As a result you were under no commitment to carry forward as a result of successfully passing your tests at Albuquerque and you will be under no commitment as to the future if you pass the tests at Pensacola.

But I think a properly organized astronaut program for women would be a fine thigg. I would like to help see it come about. It is possible that my previous experience with group efforts by women in the air can be of some value or help in connection with the possible "Space Program for Women". During World War II, first with a group of women pilots I took to England

- 2 -

for flight service there, and then later at home as head of the WASP (Women Airforce Service Pilots) it was demonstrated to the satisfaction of the authorities that women pilots were as able and serviceable as men pilots -- at least in non-combat duty because our women were never in combat duty.

There was a manpower shortage to be considered in those war days and it can hardly be said that there is any such shortage with respect to astronauts. Nevertheless the potentials of women in this new space field are worthy of determination. A group is necessary to get any sound conclusions. Therefore it has been my desire to see a sufficiently large group in the initial stage to permit of meaningful findings. The number of twelve is none too large. It might well be broadened. But it is important from the standpoint of any women's program that may develop that all who have been invited be present at Pensacola.

I have served as special consultant to Dr. Lovelace in connection with the first phase. He is the one who worked out the arrangements for the tests at Pensacola. While I will not be taking the tests I expect to be with you there for some part of your stay. In this way I can get acquainted personally with all of you and perhaps can be of some help to you as a group.

You should see that you are in top physical condition on September 18th. Take lots of exercise and if you smoke I recommend you at least cut it down greatly. We are counting on each one of the group to pass with flying colors. That's the best way to make a program for women more probable.

Some publicity might develop around these tests of women to determine their qualifications for space flights. It is recommended in the interests of a possible continuing program that between now and the end of the tests those to take the tests refrain from individually participating in such publicity. It would seem more in the interests of a possible continuing program to let such publicity as might develop deal with the group as a whole, be factual in nature and come at the end of the tests from or at least with the approval or consent of the officials who have had charge of the tests and checks.

If for any reason you wish to contact me prior to September 18th, you may write me at the above address.

Sincerely yours,

Jacqueline Cochran

P.S. This same letter is being sent to each of the women who has been invited to Pensacola.

flight, meteorology and things that have to do with design of aircraft and engines." Thank goodness these were things I already had to study in my business, for this was not a lightweight list of subjects.

Jackie seems artificially worried in a letter to Lovelace also dated September, "[Jerrie] Cobb is not on the list. Did she not pass the initial examination at your foundation and will she not be with the others at the Pensacola tests?" Jerrie had to pass the tests or we could not try!

This letter also lists ten checks Jackie sent to women for the flight to Pensacola. Two, including Cobb, received no compensation in this round. She goes on to write, "I want to see this project work well through the present and possible future stages. To be successful it will have to be held to a group effort basis. Some may need encouragement and a pat on the back. Some may need a bit of controlling. None should be treated as seconds or thirds. There should be but one yardstick." Although she is still having trouble adjusting to the importance of Jerrie, I think the letter shows that Cochran was behind us all the way.

A September letter states the women who were tested held an average of three thousand, one hundred and eleven flying hours, but is signed "W.R Lovelace II by J". It also indicates the cancellation of the Pensacola arrangements by the Navy when Lovelace's secretary closes, "...have advised girls and will advise him as soon as possible." This means I knew about the cancellation of our jet flights before our champion knew about it.

But in '61, the only time I heard from anyone connected to the testing program after leaving Albuquerque, we were going down to Pensacola to fly jets and undergo more testing. We purchased our airline tickets. Then we got the telegrams telling us the whole thing had been scratched.

WHY?

(Janey Hart) When we were trying to understand why they cancelled us out of the program, we said we'd get our own jet time. Next, they said we had to be engineers. Then they found it was possible some of the men were not, so they quickly developed some kind of GED engineering degree for them. It made me angry that they kept doing things for the men, but not for us.

The whole of NASA was in the hands of James Webb from Oklahoma who has yet to join the twentieth century. His attitude was to pat women on the head and say, "We'll take care of you little ladies."

With Phil in the Senate, I got an appointment with Vice-president

Lyndon Johnson, chairman of the space program. I was sure that if we could only get through Pensacola...but he really didn't help much. As it all came down in the end, he was probably right, because everyone else seemed to be against us.

Congressman Victor Anfuso held a hearing in the House of Representatives where Jerrie and I testified. This was when John Glenn testified against us, giving the reasons he felt we shouldn't be in the program. The one who became a diver after he had his orbit, and the one who shot the golf ball up on the moon were there with him. The three of them and Jacqueline Cochran all testified against us.

That's the one who surprised me.

Well, it doesn't surprise me. She was past space participation. Physically she couldn't do it, with her age she couldn't do it. She just didn't want somebody else to eclipse her. I spent a day with her before the hearing and I thought she was going to be more favorable. She called me to come to upstate New York and meet with her husband. But, I'm a wild-eyed liberal compared to their ultra-conservative politics. Floyd was pretty quiet, but when the cards were counted, they were stacked against us.

I think we all told Jackie she would not fit the parameters NASA had set. Which was too bad, because we know now that they just used arbitrary figures.

On the other hand, Cochran paid for our testing. So my feelings about her swing both ways. Although we didn't get into space, the testing was still worthwhile, personally and for the future. When Sally Ride's flight occurred, I was at lift-off. The mission commander looked like a college kid, but he was a general. He said on the CBS network that he regretted that our women in space hadn't happened in the sixties. He said it was foolish not to have begun then because it would have been so easy to include us.

I feel Jackie Cochran was a deep, easily misunderstood person. When I first met her, she seemed to have a swelled head. After I learned how comfortable she was in her own home, and how delicate she was on the inside, I told her I was one of the few Ninety-Nines who refused to stand when she entered the room. She wasn't surprised.

I was angry when she died, because I felt we could have become good friends. Jackie did not mislead us. She never said we would get into space. Later, when I had the chance, I told her how disappointed I was that she did not take this final opportunity to advance women, to stick up better for us in Congress instead of agreeing with Glenn. Everyone in Congress seemed worried that if anything happened with a woman on board, it would scuttle the entire space program. Like women get deader than men?

Our Mercury Physical was worthwhile, despite all the discomfort and disappointment. However, I still can't believe that NASA decided putting women into the space program would ruin our national race to the moon and stars.

They couldn't put a ladies' room and a men's room in a space capsule? "Gee, all you have to do is have a crew of women." This statement of mine didn't go over well. We were told to stop the nonsense. Politically, no one wanted to tackle the idea of women in space.

SWORN TO SECRECY

After Janey and Jerrie went to Congress and tried unsuccessfully to convince them women should be in the program, a door was closed. Disheartened and sworn to secrecy, we returned to pick up the pieces of our individual lives until we had that first reunion at IWASM. In the climate of the sixties, I don't think we could have made any more headway, since we were sworn to secrecy. If we had been together as a group, perhaps the congressional hearing would have produced swifter results, but with us scattered and busy in our own lives, there was not enough impact. Susan Schulhoff Lau raised the money to have the Mercury Women get together for a reunion when she realized that 1986 was the twenty-fifth anniversary of our tests. We are a much more tight-knit group of women now and we stand behind the women who are in the current programs at NASA.

In a quote from uncorrected proofs of *Jackie Cochran: An Autobiography* that she was writing for Bantam in 1987 with Maryann Bucknum Brinley, I read that in 1963 Gordon Cooper said, "All this talk about brains and dames in space is bunk. If there had been a scientist on my flight, I don't think we could have gotten him back. As for the ladies, to date there have been no women—and I say absolutely zero women—who have qualified to take part in our space program."

Born in a southern milltown where she grew up with an abusive foster family, Jackie had begun her flying career thirty-one years before this extraordinarily stupid statement by Cooper. By then Jackie had broken the sound barrier in a Saberjet F-86. Read Chuck Yeager's biography to learn about the stress that puts on the body. Given the same chances, at least a few of us who were younger, equally physically fit women could have met the Mercury Astronaut challenge.

When writing about NASA, Jackie is honest about her life, but responds to people like Cooper with some of her own brand of bravado. "Yes, they were discriminating, but not directly...NASA never did deny women entry at first. What they did was limit candidates to applicants who had experience as jet test pilots (and what woman except me could claim that on her resume) and to individuals with heavy engineering backgrounds and certain age brackets. The age factor eliminated me immediately.

"Women were not eliminated because they were women by any means, and I told the investigating committee so. But there was also no reason to stop the medical research on women surviving in space either.

"The committee worried that women in the program would slow down the race for space."

On August 1, 1962 Jackie sent the Mercury Women her prepared statement in the following letter.

Finally, in a draft copy of a letter I found written by Cochran to Robert C. Rurak from the *New York Telegram and Sun*, she chastises him: "What you did, Mr. Rurak, was to dwell on a part, but only a small part of the last sentence in a three page statement and to disregard in effect the rest of the statement.

"But it was an amusing story and I forgive you. Only, please don't publicize me as against women in space. They are sure to be there...I only want them to get there in the right way, at the right time... ."

Although there was a policy to "encourage women applicants" for the United States Astronaut Program by 1964, women were not actively recruited until 1976. It took us more than twenty years after the Mercury Tests to get the first American woman into space as a mission specialist (Sally Ride) and twelve more after that for one to go into space as a pilot (Eileen Collins). We are delighted that there are women in the space pro-

630 FIFTH AVENUE
NEW YORK 20, N.Y.

OFFICE OF CHAIRMAN OF THE BOARD

CABLE ADDRESS: JACOCHRAN, NEW YORK

August 1, 1962

Mrs. Bernice Steadman
Trimble Aviation
Bishop Airport
Flint, Michigan.

Dear Mrs. Steadman:

As you probably know, there were some hearings before a House Committee of Congress last week on the subject of whether women have been or are being discriminated against in the astronaut program.

Miss Cobb and Mrs. Hart testified and also I testified. There was testimony by a representative of the NASA and Astronauts Glenn and Carpenter.

You already know my general views from my letter to Miss Cobb. A copy of my prepared statement is enclosed. When it comes out I'll send you a copy of the Congressional Record containing all the testimony.

Also I am enclosing copy of my letter to the Counsel for the Committee which I believe is self-explanatory.

I spoke only for myself without claim to represent any of the women who passed the Lovelace tests. Miss Cobb stated she was the spokesman for all of the 10 not present. I am satisfied this is not true as to some who had already expressed their views to me. Was she authorized to act as spokesman as to you?

I'll be glad to hear from you.

Best wishes.

Sincerely,

JACQUELINE COCHRAN.

---

gram now who are shuttle pilots, who are shuttle commander quality. That's what we all were after in the Mercury program. None of us, men or women, were scientists. We were just very excited pilots.

1963 International Race

Solo flying is a challenge of courage. It is a precious time alone when the only pilot you can depend on is yourself. But if a pilot wants to win, I do not believe a transcontinental or an international race is the place to fly solo, unless that is defined as a part of the overall challenge for everyone.

On May 29, 1963, Mary Clark and I started in fifth place in the All Women's International Air Race (AWIAR) but whipped past the other planes as the first plane to arrive in Detroit. This race started in Welland, Ontario, and ended in Hollywood, Florida. Stopovers were in Michigan, Ohio, Tennessee, Georgia and in northern Florida. This was my first international race after the Mercury tests.

In Canada's wild country with miles and miles of forests, the airports don't stand out well from the rest of the openings in the landscape. In any case, clear weather or foul, we flew low as much as possible so we didn't have much forward visibility. There seems to be little effort to line roads up N/S, E/W with the compass in Canada. I couldn't pinpoint airports electronically from a distance with the equipment we had then. Most of the Canadian homing stations were not on an airport because they were departure VORs, located where the pilot transfers to radar control. With all these challenges, it was nice to beat everyone else and go through customs in Detroit with a feeling of success.

In addition to the navigation challenge in Canada, the weather deteriorated. When we landed in Albany, Georgia, for a rest stop, some of the girls had a cup of coffee, but I immediately looked over the weather reports. ("Oh, boy! We've got to go, because we've got to beat this front.") We flew out just in the nick of time, on the good side of the wind. We were racing in a Piper Comanche. With one hundred and eighty horsepower and a handicap of one hundred and fifty, we had to fly faster than one hundred and fifty miles an hour to earn any of the points that would win the race. The Comanche was first designed for a bigger engine, but that engine wasn't ready. Fortunately, our airframe was rated for a lot more action than we would give it. We went up on top to fly over some building-cumulus. We could still see the ground through openings in the dark and ominous clouds, but those who stayed underneath were having trouble. Some of them called, lost in the vicinity of Gainesville. This race emphasized that weather is always a matter of timing, especially with cumulus. They are either building or diminishing, never stagnant.

Rules and Choices

We always raced under Visual Flight Rules (VFR) for daytime. It could be VFR-on-top, as we flew in 1963, only if our ascent and descent were flown with the visibility and spacing of actual VFR. We wouldn't be disqualified if we wound up having to get through weather. The problem is the fine line between a controlled VFR approach and one requiring the more restricted Instrument Flight Rules (IFR). Circumstances, like the heat rising from a city, might open holes in the overcast here, but not over there. To the best of our ability, we tried to adhere to the rules, mainly because that was the safest flying for all of us.

There was a designated takeoff time and a designated day and time for the race to finish. We had to land before official sunset each day. If we came in with a clocked time past the time published in the official's handbook it was grounds for disqualification, even if someone gave us the wrong time when we were in the air.

It is not likely that a lengthy race will be flown without encountering some weather problem. We'd fly at the crack of dawn, quit in the middle of the day, and fly again in the evening to try to minimize adverse winds and turbulence. Headwinds were a real drag, but tailwinds never seemed to overcome the time we used to climb.

To get ready for any race, I would gather all the climatological data from the prior ten years that affected each planned stop along our route. With this, I could plot an idea of the average surface winds and winds aloft on a wall map. I could then choose the route I wanted to fly. With each leg pre-plotted, we knew the miles and the time it should take to fly the race. We knew the elevations all along the route. We knew how much time it was going to take to climb to altitude and when we should descend to land. My co-pilot and I tried to hold to that plan for each race as much as was possible under actual weather conditions.

I continued to check weather on the route myself. Now I believe most pilots call a private weather service. Those services take everything off a machine. I don't know how I'd do under the new circumstances. When I was racing, pilots could get to the local weather bureau and look at the charts. I would look at a variety of their weather maps, see the daily upper air data, and then talk with local weather-wise people. It was a lot easier for me to do it than to try to explain what I needed to somebody two thousand miles away. It is hard to make someone not in the races understand all the factors in making a decision. They might make a recommendation based on being first across the finish line rather than shortest time in the air. I might stop and choose to wait a bit to gain best winds, so I didn't care who crossed the finish line ahead of me as long as I won the race!

Mary's family business, John Crowley Boiler Works, and my Trimble Aviation sponsored the plane in 1963, so we didn't have to cover the racing expenses by winning. I've just realized that in all my racing, I never really went out to look for a sponsor. Lucille Quamby had a Brunswick sponsorship because she had a contact who was a top bowler and representative for the company. Janey Hart sponsored me in several races. When I had Trimble Aviation sponsor me, it was good publicity. My business sometimes paid the racing expenses. I put up the airplane and fuel expenses. I think we always took care of our own personal expenses.

A prize of a thousand dollars was not enough to cover the cost of anything. Honest to God, I think I would have raced for the trophy alone. Almost every day I was instructing from sunup to sundown. Racing was a welcome change, a way to express myself and my own ability. It gave me the opportunity to meet other women doing similar things and doing them very well. I matched my talent against theirs to find out where I stood. I not only enjoyed the competition, I enjoyed the people with whom I was competing. These women thought the way I did. We respected our equipment the same. We each did a lot of work to get to the point of racing, and we each wanted to win.

To prepare for winning, I would practice timing my take-off and landing speeds before a race. I used to play as much tennis as possible. Going through the hottest parts of the United States in the heat of the year, we had to be in good physical shape. The airplane was inspected very specially with a focus on all the parts that might be susceptible to the strain of racing. We didn't want the drag of open cowl flaps to add even a few seconds to our total time. The prop and governors had to be in good shape. If the engine needed an overhaul, I made sure the insides of the cylinders were as friction-free as possible. I never had the kind of money to do what Edna Gardner Whyte did with chrome cylinders and that sort of thing, but there was so much room for human error, our judgment compensated sometimes for not having the chromed cylinders.

I figured flying the airplane and having the course all laid out were pretty much the job of myself as PIC. However, I went over everything with my co-pilot (Second In Command) so she knew where we planned to go as well as I did. The terrain might have been new to her, but it might have been new to me too, especially during the Internationals.

Each racer did her own calculation and planning. Our altitudes were not assigned. Finding the best wind during a race was by trial and error. Winds aloft reports aren't always that accurate. Taken every twelve or twenty-four hours, what I experienced might not have been what they reported or forecast. Therefore, once in the air I would arbitrarily pick an altitude, depending on where I felt the best winds would be and would keep in mind how much time it was going to take to get to our destination. When I would get to that altitude, I'd go up as much as a thousand feet higher. After I leveled off, my co-pilot immediately helped run a ground speed check. If I wasn't satisfied, she rechecked as we stepped down to try to find a better wind. I never went much below my planned altitude, however. To reclimb later along a route might be too great a loss of time. I always tried to make up for the total time lost in climbing by planning when to start our descent near the destination. Then, with a slight nose down attitude, we would pick up speed, penetrating the air more easily.

We were required to land at a designated airport every time the route took a real change of direction. Between these, faster planes might fly over stops where slower planes had to refuel. Turns in the route and required stops prevented cutting corners and created a somewhat more equivalent race between widely different planes. Designated stops for everyone were spaced about three hundred miles apart all across the country, so everyone was climbing and descending a number of times in the race.

Experienced pilots could save a great deal of time by planning how to balance these stops.

When we saw each other at the stops, the truth was only something on which to build. "We've got plus twelve points" (meaning twelve miles an hour over their handicap) could be an absolute exaggeration of what they really had done. But all of that was part of the fun. I don't think anyone knew how anyone other than themselves really did until the scores were out. Then, it really was fun to win.

Mary and B admire their 1963 winnings.

Ground Effect

During another International with Mary, going into Nassau, I flew the plane so low we were getting sea spray on the windshield. Flying as low as possible over the ocean, it seemed like we were flying in a wind tunnel experiment, showing how the smoke pours off a model. We were flying in ground effect, where a "v" of air gets compressed close to a surface. When you're in that compressed air, it's just as though somebody has grabbed ahold of the seat of your trousers and is shoving you forward. But ground effect is only as high as the width of the wing of your plane. Get even a little above that, and this wondrous effect bleeds right off. In this race, we flew in ground effect just above the waves from the mainland to

Nassau. You can be certain that I was concentrating on instruments all the way. With almost no altitude, it would have been very easy to fly right into the water. Mary suddenly yelled, "Oh, look! There's a whole bunch of fish down here... Wow! Look at that fish!" I did *not* look.

When Very High Frequency (VHF) navigation became available, we still had no navigation aides when we flew low until we were pretty close to our destination. I'd weigh the risk of getting lost against the possible gain in time. Flying long distance, it doesn't take very many degrees to be off by quite a few miles. Flying across the ocean without any landmarks carries an increased risk. Pilots have to weigh each choice they make against the odds and take only the risks they think they can comfortably handle. One of my choices was always to have a co-pilot who was as aware of the developing situation as I was.

WACOA, with Janey Hart

The idea for the Women's Advisory Committee on Aviation (WACOA) was established during the Kennedy administration. Thirty-two women were to be selected representing as nearly as possible a cross section of all aspects of non-military aviation in the United States. When things calmed a bit after the sudden, unexpected transfer of the presidency required by Kennedy's assassination, Janey Hart was asked by President Johnson to handpick women for the first Committee.

Short Notice
(Janey Hart) What I remember was Liz Carpenter, LadyBird Johnson's press secretary, calling. I asked when they would like to have this group start work. Carpenter gave me about ten days! I mean it was short notice. Fortunately, Alice Hammond and Jean Howard were in Washington, so I called them over and we immediately got down to work. We created a list of people who covered a range in geography and skills. We wanted input from as many areas as possible. I sent telegrams.

Janey went through the files of the FAA and came up with a marvelous cross-section. There were women from the airlines, general aviation, education, medicine; just about every aspect of aviation. We were to work directly with the administrator of the FAA and various echelons under his jurisdiction. Information was to come from them about what they were going to do, and we were to feed back to them what professional people in the field felt should be done.

The President wanted us to study the FAA, its projects and actions, to develop ideas and make suggestions for more efficient aviation operations. We were to suggest things that weren't being done that we felt should be done. We divided the group into subcommittees to study specific questions. Then I found a good place to meet, where we could work hard and yet relax.

Our first meeting was held in the FAA administrator's conference room. My impression of that room was that it looked like the war room from the movie, *Dr. Strangelove*. We sat at a big round table, each with our own microphone, in chairs so heavy we could hardly move them. We were asked to say a few words about where we thought the Committee could make a difference and what we thought about the assignment. I said I thought it was a bit of tokenism; that they probably wouldn't listen to us, that we were simply a publicity gag. I also thought there was a lot to be learned on both sides if we could really get on even ground.

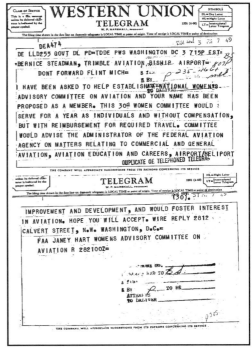

Telegram from Janey.

About an hour later the funniest thing happened. The Administrator came and addressed himself quite specifically to our remarks. It became obvious that we were being recorded, so we began to say "turn off the recorders" anytime we didn't want to be overheard and put our water glasses over the microphones. Isn't that crazy?

The administrator when I was part of WACOA was Najeeb Halaby. Mr. Halaby was an imposing man who always seemed sure in his mind what the FAA needed or did not need. Sometimes it was difficult to get him to see things our way, but it was fun trying.

It took a lot of time for some of our recommendations to get into place, but almost everything we suggested was eventually done, and as recommended. Bernice's big project was upgrading instructors with the equivalent of a master's degree in teaching. They certainly have upgraded that. The Committee also felt there ought to be parallel runways for general aviation at the big airports, instead of having jets running up your rear, and they have done that almost everywhere, at least where there is a new airport.

We had a constant fight with the bureaucracy. Because I lived in Washington, when I found things coming apart in a quiet corner away from the direct gaze of the nation, I could call Bernice and say, "They're trying to tie it up in red tape. They don't even understand what we are saying." She'd fly out, we'd sit down in a quiet corner with whatever group of people was causing the holdup, and they'd discover... "Right, that's what we want to do, too."

Ninety-seven percent of what we recommended has been acted upon.

PHILIP A. HART
MICHIGAN

COMMITTEES:
COMMERCE
JUDICIARY

United States Senate
WASHINGTON, D.C.

May 21, 1964

Dear Bernice:

It was with more than casual interest that I read of your appointment to the FAA women's advisory committee.

Naturally, I was quite pleased to learn that Michigan has apparently cornered the market in outstanding women fliers.

Sincere, though belated, congratulations. Your past record would indicate that such recognition is long past due.

Best wishes for continued success.

Best regards to Bob.

Sincerely,

Philip A. Hart

Mrs. Robert A. Steadman
12214 North McKinley Road
Montrose, Michigan

Airman's Subcommittee

I became the chairman of the Airman's Subcommittee for WACOA. Our focus was on the fixed-base operational side of instruction in aviation: pilot techniques, attitudes, and skills. One of the things this committee was determined to develop was a Gold Seal Certificate to separate dedicated flight instructors from the ones who were just using flight instructing to build up time to go into the airlines. We hoped to give Gold Seal instructors additional responsibilities. We set up the program so the FAA would not have to increase their personnel because additional

---

TRIMBLE AVIATION
Incorporated

August 6, 1964

TO: Jane B. Hart, Chairman,
Women's Advisory Committee on Aviation
All Committee Members

FROM: Airman Proficiency and Performance Committee

RE: Recommended action by Committee and Flight Training Program Proposal.

After much deliberation, discussion and thought the "Airmans Proficiency and Performance Committee" would like to submit to the Committee as a whole a proposal for an incentive approach to the pilot shortage. We enlist your thoughts, criticism and suggestions. If we have a positive indication we would like to submit it to F.A.A. or other Government agency for further consideration.

Please address your comments to:

Bernice T. Steadman
Trimble Aviation
Bishop Airport
Flint, Michigan

Sincerely,

Bernice T. Steadman

inspections would be in the hands of people who understood instructive flying. We wanted to reward the people who really cared about their students. Halaby felt there was no difference in instructors. He felt that if pilots could pass the FAA exams, they were good instructors. We said there were good, bad and indifferent instructors who were passing the exams. Our recommendation met with great resistance, but ultimately, we did get a watered-down version of the program into place.

The government system seems to be a pyramid: the more people you have under you, the better off you are. Being mere taxpayers, we didn't understand we were removing their need to hire more inspectors. We finally addressed ourselves to what we could get through the bureaucratic channels. Our recommendation was to allow Gold Seal instructors to give ratings; they could give flight tests instead of making students await the FAA's pleasure.

Another thing I was still interested in was the student dropout situation. If bad weather interrupted appointments two days in a row, the student often quit. I wanted time in flight simulators that had the look and the same feel as the training airplanes to be an optional percentage of the time toward the total requirement for a private license. That way, no matter what the weather was, students could train. They would be in a comfortable, non-threatening environment and could at least learn radio, navigation, and instruments.

> When the FAA said they were already investigating this idea, we flew down to NAFEC (National Aviation Facility, Experimental Center) to look at their simulators. There was quite a layer of dust on them. One of the girls put her initial or something in the dust. We went back later, I mean much later, six months later, and there was the writing in the dust. So they weren't doing a damn thing. We made the recommendation that FAA close NAFEC. Wow! The FAA went into orbit. "What do you mean you want us to close it down? What's the matter with you?"
>
> We reported their facility was a waste of money. "All you're doing is throwing chickens through engines to see what effect they will have." I mean one after the other, chicken after chicken, thrown through jet engines. It seemed like one chicken would give them the answer! We got them all shook up, but not closed down.
>
> WACOA rotated people from time to time, including chairmen, so eventually there was a complete turnover. Committee meetings took a weekend

plus one day on either side. We worked hard, without a lot of playing. Each sub-committee hashed out recommendations during the year. After this, the Committee met as a whole, then the FAA administrator came with his assistants from the area we were working on, and we presented our findings. WACOA served for more than one president. I was rotated off by Nixon's term.

Politics Again

This was serious stuff for us and for aviation. The Committee was set up on a year-to-year basis. Janey and I were on WACOA the same length of time.

So many things in politics are tied to a political party. I tried to avoid that, both personally and for another reason: If we had gotten into political diversions, we would not have accomplished much at all. As we worked we really were dealing mostly with civil service career people and the bureaucracy. At that time, if they had been working a long time for the FAA, most of them were Democrats. Still, there was good representation from the Republican side of the aisle.

I put one person on the committee who continued as an FAA advisor. It was our balloon friend, Connie Wolf. She was a very conservative person. Her husband was a legal advisor for the Aircraft Owners and Pilots Association (AOPA), an organization with considerable power politically. AOPA could have made difficulties for the entire process. The two of us became good friends, but I always knew we would have a negative reaction if she didn't get the information before the issue was discussed. So I would sit down with her to get her all fired up on the issues. Thus, AOPA never gave us a problem.

Good thing for WACOA that you had that perception. My Airman's Sub-committee won the application of ten hours of simulator time counting toward a private license. We also changed some of the medical requirements. One crazy thing we ran into was that if you had an overbite you could flunk your physical. I knew this was antiquated nonsense because I had one. "What is this? Why the overbite? If you can chew your food, what is the problem?" When they did the research, they found this rule came from the civil war days when a soldier had to be able to tie off a powder bag with his teeth. Really! I'm not kidding.

In the end, the FAA began to listen to our Committee with both ears and AOPA, with all its money, was right there with us in each campaign. Janey and I were serving on WACOA when they came up with the wedding cake thing: a step-down of controlled airspace into a large airport. We told them that without preserving unrestricted lower air spaces near airports, where commercial carriers did not need them, the little guy in the Piper Cub from a grass strip would have nowhere to fly.

WACOA was a very interesting thirty-two-woman group. One of the women was Dr. Helen Brown, Amelia Earhart's personal physician. A couple of them were also with the FAA in responsible positions. Several were with the airlines, and some owned fixed-base operations like I did. It was a great experience to work with everybody. As a member of this Committee, I got better acquainted with Page Shamburger. We became fast friends and worked on numerous projects together for the Ninety-Nines, Inc.

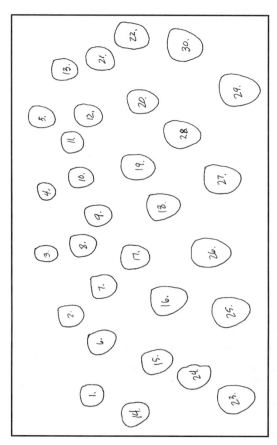

WACOA circa 1964. (Left to right, with apologies for my memory. You will recognize many of these women as people who helped create my story). 1. Gini Richardson 2. Dr. Dora Dougherty Strother 3. Gene Nora Jessen 4. Jackie Joo 5. Connie Wolf 6. Jean Pearson 7. B Trimble Steadman 8. Louise Hyde 9. Jerrie Cobb 10. Janey Hart 11. ? 12. ? 13. Louise Smith 14. Charlotte Kelly 15. Katherine Stinson 16. Alice Hammond 17. Betty Miller 18. ? 19. Blanche Noyes 20. Nona Quarles 21. Jimmie Kolp 22. Jean Ross Howard 23. Page Shamburger 24. Fran Noldie 25. Dottie Young 26. Betty Gilles 27. Jerrie Mock 28. ? 29. Virginia Britt 30. Katherine Hiller. Names

also on the roster were (some women were not in the picture): Mildred Alford, Nancy Livingston, Eunice Naylor, Margaret Boylan, Julia Short, Billie Trimm. I'm sorry I can't sort them all out. Each held impressive credentials.

Racing: 1966 Transcontinental

In transcontinental racing, I generally flew from Michigan to the closest race point to fly the course in reverse out to the start of the race. This helped me pick up some landmarks that would come in handy on our return flight, like the time we had to reroute during that first transcontinental race. I'm still amazed when I remember how fast that particular desert thunderstorm grew. It certainly taught me that when we were circumnavigating weather, my co-pilot was to watch closely so we didn't wind up having to make a ninety-degree turn, or worse, to get back on course. In that first race, if it weren't for the one place I remembered in the landscape, we would have lost so much time figuring out how to continue, we would have placed nowhere near the top ten. Thank goodness for rabbit-eared mountains!

I have yet to find a natural pilot. I've found people who take to it very rapidly, but I think flying performance is a learned response. Flying is too complex to be something we are born with. Racing takes a lot of practice. Many pilots also do special things to groom their airplanes. They don't change the stock qualifications, but they do everything they can to make their machine perform as a single, smooth unit. They also develop their own secret formulas, special instructions for leaving an airport, getting to altitude and settling on course.

The pylon racing I did on weekends when I was a student in Flint was a small thing, but was actually real competition. Around the pylons, the planes got awfully close together and close to the ground. As I look back at the four to six planes racing at one time, there were too many of us racing. The flight instructor who flew a Luscombe used to just tear away until he got to the pylons, then I'd pass him. In long distance racing,

I was mighty glad I knew how to fly a pylon turn without my plane using up any extra time and space as it slipped through the air.

Weather

Weather bureaus were very busy just before and during a race. It was often difficult for contestants to get up close to the sequence reports that told how the weather had changed over the latest period of time. I think this crush at flight service stations during a race is what caused many of the pilots to go to their own weathermen. However, weather is such an important part of flying, I wanted to keep control over my choices. I also enjoyed the challenge of following the most changeable piece of the racing puzzle. A certain amount of book learning and a certain amount of my own experience with weather gave me some wisdom in terms of what I wanted to fly in, and especially what I *didn't* want to fly in. However, even the most experienced weather prognosticator misses once in a while. Airmass weather can catch any pilot. But frontal systems are a little more controlled, so it's a little easier to predict what's going to happen in front of them. I learned I could cross over the tops of some fronts and not others. When I did my own checking, I felt a little more confident because I understood the basis for making each decision.

The AWTAR contestants were strung out so much, flying over a three and a half-day period, we might or might not be flying in the same air mass as the others. It depended on how we each timed our flight. Several races were won by the people who held back. Other times, we got to the finish with the rest still back there, somewhere. It depended on what weather was really like as each pilot made her decisions.

When we stopped at designated airports, the time on the ground was not counted against us. When the weather got bad enroute, if we landed at an airport that was not a designated airport, we might just as well quit, because the rest could go on with the race. If we had to stay overnight at one of these unscheduled stops, there would be no way that we could place, unless a big enough group had to land for safety's sake. In that case, the officials would lengthen the time of the race. The chance was, if I went on fighting weather, and most of the others stopped, getting much better weather the next day, my time in the air would not be good enough to win. A pilot either likes that kind of competitive flying or doesn't get into it at all because it is *hard*. In the 1997 history of the Ninety-Nines, it states I raced in twelve Powderpuffs and was in the top ten in all of them. That was a lot of planning, a lot of intense work. I never took my hands off the controls in a race. We were too busy, using every means we could imagine to improve our performance.

Two Weeks

A race was a two-week event by the time we went out to a start, turned around and raced back. The AWTARs were more tightly organized than the early Internationals. There were specified times to do this, that and the other thing before an AWTAR. Before we took off every morning at an International race, I knew there was a morning pilots' meeting somewhere, sometime... and usually everybody was there at the right time! At 8 o'clock, I knew I had to be at the pilots' meeting for the AWTAR. International pilots could stop at any designated airport, even if they were not required stops. We clocked different times on different routes. The AWTARs had designated stops to try to equalize the difference between planes. These were must-stops, so everybody took basically the same route. There were leg prizes, so there were races within the race. Even if we knew we wouldn't win the overall race, we were always racing each leg. Sometimes I won recognition by studying the route and finding a small advantage.

At any of these racing stops, the priority to land was given based on the proximity of the plane to the end of the runway when the first call was made to the control tower, unless someone was low on gas. As we got close to any destination, we could tell where other planes were only by listening carefully to the calls. If we gave way to a pilot in trouble, officials would note that on our card so our time could be adjusted. Everyone appreciated this safety feature, because we gave away no advantage. The potential for change in protocol forced both pilots in a race plane to remain alert all the way down to the ground. During the later races, after a pilot landed at a designated airport, she stopped her plane as close as possible to the officials, adjusted the mixture to keep the engines going and stuck her ticket out the window for the official to stamp. It still was a terribly focused moment.

Competitors

Then there were competitors trying to do better than..., but we could never see that going on between any two pilots. In my own case, I wanted to beat Fran Bera because I thought she was the best racing pilot going. Fran was a mentor of mine before she moved to California. She led me into racing. She also did such an efficient job of plotting, working all year to get her airplane ready, she made me work hard to win. Then Fran got to the point where it seemed she was flying almost anything. One year she raced an airplane we weren't too sure was going to hang together. It was in terrible shape! But she placed very well even in that race.

The 1966 AWTAR was from Seattle, Washington, to Clearwater, Florida. My husband and I were living in Montrose on "the ranch." Trimble Aviation sponsored my Piper Comanche. Knowing the limits of my airplane, with the ten-year average for climatological data, we plotted the race route so I would make as few stops as possible. As racers we had to fly the prescribed routing, but we could skip stops in the faster planes as long as we landed at all the must-stops, where the course changed direction.

We hung a large map of the contiguous United States across one wall in the house. Plotting on the map, with pins in each stop, we took a long piece of string and measured the difference in distances using various combinations of stops, focused on the must stops. We finally found the route that was the shortest for each leg, with one difference, where the deviation would be faster than flying the most direct route. Then the direction of the wind and the weather had to cooperate with my planning.

This race allowed about four and a half days, always getting up before daylight, checking the weather, making sure the airplane was serviced, going over the charts. Then the nights of the race were also busy: the plane was polished, Mary Clark and I made sure we knew the NOTAMS (Notices To AirMen) for the next day and we made sure the radios were working right. Once a Piper dealer took a brand new radio from his plane and put it in our airplane when we hadn't the time to get ours fixed. I flew the radio back to him after the race. By the time we were done with our plane and the next day's planning, we were tired. We always knew the next morning would seem cold and damp. We'd go in for breakfast, predicting green eggs and ham again. Theodore Seuss Geisel may have adopted our awful mess as the title for his book. Another famous character, Charles Schultz, flew with his wife in an AWTAR. They gave everyone a special Peanuts pin, created for that race.

My co-pilot made sure I did what I said I was going to do. In all of the races, I tried to stay exactly on course. A mile off course, we knew we couldn't win the race. If the timing was off, we immediately reviewed what the reasons might be and adjusted our flight accordingly. Each time I finished a race I felt I had learned from mistakes I'd never do again. Well, I didn't, but I would do something else! There just didn't seem to be any end to the lessons. The 1966 AWTAR turned out to be the one race I was able to fly exactly the way I planned it.

WEST TO EAST

Flying west to east, we always carried oxygen. On the first leg in 1966, the weather was bad, with clouds hanging on the hills. Mary and I flew the first leg from Seattle to Portland to reach the only route that we could use to cut off some of the required climb time to get through the mountains. By flying this route, we also burned off some fuel down low, so our climb was faster than it would have been right at the beginning of the race. All the way down the coast to the Columbia River we were low to the water. Turning east after contact with the Portland airport, we

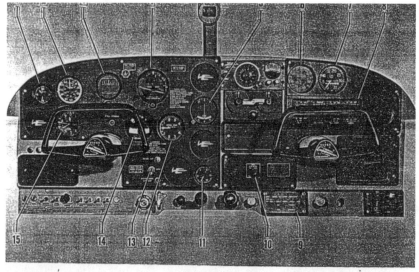

1. CLOCK
2. AIRSPEED
3. DIRECTIONAL GYRO INDICATOR
4. GYRO HORIZON INDICATOR
5. VOR INDICATOR
6. MANIFOLD PRESSURE INDICATOR
7. RPM INDICATOR
8. INSTRUMENT CLUSTER
9. RADIO CONTROL PANEL
10. AUTO CONTROL KNOB
11. SUCTION GAUGE
12. RATE OF CLIMB
13. LANDING GEAR
14. TURN AND BANK INDICATOR
15. ALTIMETER

The Comanche panel.

climbed quickly as we passed Mt. Hood. Above the haze, I looked down at the mountain, and watched an avalanche! It was in sight long enough that we took pictures of it. Then we continued to fly at the new height for quite a distance, because the winds were favorable.

After we crossed all of the western mountains, the route I laid out sent us flying low over water and swamps the rest of the way to Florida. We stayed close enough to civilization to be able to get to dry land in any sort of an emergency, but to walk away from landing in the slop below us wasn't nice to contemplate.

At the end of the race, Clearwater was hard to find because we were flying so low. If I remember correctly, we had our eyeballs on the windshield trying to find the airport. Everything looked the same. Every jut of land looked like the next little jut, and the bays all looked like they were cut with pinking shears. To finally touch down on the right pinking was a delight.

Thirty Seconds

After any of the races, there would be a series of formal programs to follow. The first day was a chance for the racers to rest while the officials did the scoring. As the pilots relaxed and played, all of the racing planes were inspected to make sure they were still stock, still safe to fly home. In 1966, when Fran Bera and I were in the pool, swimming, we were not sure which one of us was going to have to drown the other, because we knew we had both done quite well. Glad she didn't drown me!

The following night was the banquet. We beat Fran Bera by maybe thirty seconds. Wasn't that amazing? The 1966 AWTAR was the longest course we had for an AWTAR, over twenty-seven hundred miles. Twenty-seven hundred miles with less than a thirty-second difference between two planes, two pilots? Fran and I didn't fly the same time of day, didn't make the same stops. We flew identical airplanes. She flew solo and I had Mary as co-pilot. That may have made the difference. Mary Clark and I did that race right to the book, following six months of planning, and the ten-year average of climatological data was marvelously reliable for once.

This win was especially sweet for me, because Fran was still the pilot who set the standard for all of us. She won seven of the twelve AWTARs that I flew. After each race, but especially this one, I went home to restart the grind, refreshed somehow by all that work, glowing with the extra awards.

1966
All-Woman Transcontinental Air Race

Best Score in the
Piper Comanche 250/260
Class
Awarded to

Bernice T. Steadman
Mary E. Clark

for outstanding performance in the 1966 "Powder Puff Derby." This aircraft and crew achievement further enhances public acceptance of private aviation and continues to prove the unexcelled reliability of this class of aircraft for business and personal transportation.

Twentieth Annual AWTAR — July 2 to July 5, 1966

Seattle, Washington — 2765.67 statute miles — Clearwater, Florida

presented by
American Aviation Publications, Inc.,
and the editors of
American Aviation Magazine

Notes from the 1997 Ninety-Nines' History Book:
1966 Seattle, Washington, to Clearwater, Florida—longest AWTAR.
2876 statute miles   91 airplanes   165 pilots
1st place B Steadman and Mary Clark
22.71 points   Piper Comanche   208.71 mph
B Steadman set a speed record that stood for a number of years.

1947 – 1966
Powder Puff Derby
20th ANNIVERSARY
SEATTLE, WASH
CLEARWATER, FLA
JULY 2-5, 1966

AWARD OF MERIT

Presented to

Bernice T. Steadman
First Place

In recognition of outstanding participation in the 20th Anniversary All Woman Transcontinental Air Race. A 2766 mile course originating in Seattle, Washington and terminating in Clearwater, Florida during the period July 2 to 5, 1966.

July 7, 1966

Ray A. Brick
Chairman, AWTAR, Inc.

Evelyn Bryan Johnson
Chief Timer

The winner!

EXPENSE REPORT 1966 AWTAR by Mary Clark

Jackson MI	$12.23	
Louisville KY	9.43	
Memphis TN	10.58	
Alexandra LA	11.83	
Houston TX	8.10	18 gal
Harigen TX	14.99	33.3 gal
Monterey CA	5.00	
Alice TX	5.72	13 gal
Waco TX	18.45	41 gal
Laurel LA	16.45	
Tallahassee FL	14.16	
Rome GA	<u>12.76</u>	$139.79 gasoline
Landing Fee	$ 2.00	
Oil	1.98	
Hangar	2.50	
Oil Change	7.43	
Hanger	4.00	
Tiedown	<u>1.00</u>	$18.86 maintenance
Food	$132.87	
Lodging	165.91	
Miscellaneous	58.36	
Car Rental	<u>16.32</u>	$373.46 non-airplane

188 ⁂ *Tethered Mercury*

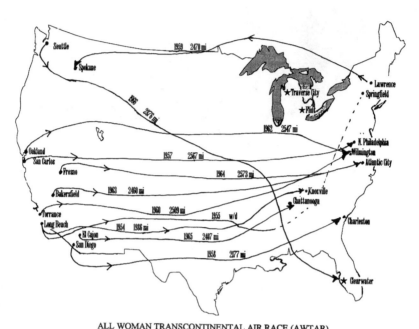

ALL WOMAN TRANSCONTINENTAL AIR RACE (AWTAR)

1954 Long Beach CA to Knoxville TN
51 planes   shortest route   95 pilots
co-pilot, Lois Wilson

1955 Long Beach CA to Springfield MA
47 planes   w/drew for emergency

1957 San Carlos CA to North Philadelphia PA
49 planes   89 pilots
co-pilot, Mary Clark   7th place
Beech Bonanza   17.482 points*

1958 San Diago CA to Charleston NC
69 planes   129 pilots
co-pilot, Mary Clark   8th place
Beech Bonanza   10.689 points

1959 Lawrence, Maine to Spokane WA
66 planes   129 pilots
co-pilot, Jane Hart   10th place
Beech Bonanza   -1.4 points

1960 Torrence CA to Wilmington DE
85 planes   157 pilots
co-pilot, Lucille Quamby   3rd place
Cessna 172   13.95 points

1962 Oakland CA to Wilmington DE
54 planes   99 pilots
co-pilot, Mary Clark   10th place
Piper Comanche 21 points

1963 Bakersfield CA to Atlantic City NJ
47 planes   84 pilots
co-pilot, Mary Clark   7th place
Piper Comanche 17.96 points

1964 Fresno CA to Atlantic City NJ
61 planes   111 pilots
co-pilot, Mary Clark   6th place
Piper Cherokee   19.56 points

1965 El Cajon CA to Chatanooga TN
79 planes   148 pilots
co-pilot, Mary Clark   9th place
Piper Comanche 13.70 points

1966 Seattle WA to Clearwater FL
91 planes   longest race   165 pilots
co-pilot, Mary Clark   **1st place**
Piper Comanche 22.71 points

*miles per hour over the handicap

Family Snap Shots

It Gets Cold Up North

A hardy group of Scotch, English and Finnish settled the Upper Peninsula (UP) of Michigan, with Irish and Italians mixed in for good measure. They all were people from lands where it was even harder to make a living than in the Michigan wilderness. My great-grandfather Trimble came as a bound-boy into Canada. This means he sold his services for ten years to help keep his family alive back in Ireland. One time I went through Flesherton, near Owen Sound in Ontario, and saw that there are a lot of Trimbles in that cemetery. My Aunt Ann who lived in the UP near Cedarville had a book showing that my great-grandfather must have had a good master, because he was taught a beautiful handwriting. His son, my grandfather Trimble, was sent to Toronto to learn watch repair. Later, when he inherited two thousand dollars from his mother, Grandfather moved to the UP, somehow met my grandmother in Sault Sainte Marie (The Soo) and bought a huge house near Rudyard. There he established a reputation as someone who was very inventive.

At the time my mother's Christie ancestors arrived in the UP, it was not a luxury to have a large family, it was a necessity. Her parents had twelve children of their own. Her father, Robert, arrived in this country during the time of the Homestead Act. He and his brother homesteaded two parcels of land in the wilds of upper Michigan near Dafter. Grandfather Christie did a good job making his crop pay but his brother was having financial difficulties. His brother felt my grandfather had cheated him out of the better land since Grandfather had come ahead to select the two parcels. Even though Grandfather had no debt, he did not want animosity, so he simply traded farms and took on his brother's debts. Again, he proved to be the better businessman and farmer. He and Grandmother Christie raised all twelve of their children and then, when his brother's wife died, they raised his two children. Later, when they were in their seventies, a son's wife died and they raised those two children until the son remarried. These children all got along very well, did the jobs assigned to them and took this family orientation with them when they left home to start successful marriages and families of their own.

I can still remember going to see my grandmother Christie. She always had a pot of tea on the back of the wood stove, which she started in the morning. In the evening, I would pour myself a cup, add cream and sugar and sip what I thought was going to be coffee. Time after time I could hardly choke it down. It was the sweetest, syrupy, most awful stuff I've ever tasted!

I was happy when the entire family would gather around the kitchen table after clearing it from supper. I'd sit among them and listen to stories of runaway horses, the pranks and the games they would play as children, and choke down her awful tea.

Dafter, and Rudyard, where my parents lived after their marriage, were two small towns that grew up along the railroad when logging people moved into the UP. Train engines were coal-fired then and required more water and fuel than was efficient for the trains to carry. About every twenty miles along the track small towns would materialize to service the needs of the railroad. Father was the stationmaster in Rudyard, a few miles south of Sault Sainte Marie (the Soo). By the time I was born, my birthplace had grown to serve as a central point for grains and other agricultural products being shipped to market in the Lower Peninsula.

My parents, Harry and Laura Trimble, had a boy, three girls, then me. Everything was going along fine when one night, during a nasty storm, a fire broke out at our house. I was only a year and a half-old, ill, sleeping with my parents. Mother says my crying woke them. They suddenly realized what was happening to the house. Father got Mother and me out safely, then went back for the rest of the children. He reached the stairway, it collapsed, and that was the end of most of my family.

This was extremely traumatic for my mother. She never really quite got over it. I was so young, I didn't realize anything except I was not in my own home. I was bounced back and forth among the relatives for a while, but they gave me loving care. Mother went out to Mayo Clinic. When she came home, she had a Lindy Doll for me and a book about the first solo flight over the Atlantic.

Soon after her return, we moved downstate to live with my Aunt Margaret in Pontiac. A man named Ray Whipple lived across the street. Eventually mother and he married and had a son. There were seven years between her new child and me. This golden-haired boy required quite an adjustment from me. His parents wanted him to have every opportunity, but what I was capable of doing was more than he could handle. I wanted an accordion. He got the accordion and the lessons. I watched his lessons and learned to use his instrument, playing well, though my brother never mastered the instrument. Later I was in the all-state orchestra, using a rented viola that I loved to play.

Pensive little miss.

There was a lot of friction between us when he was little, but my brother and I became a fine pair of friends eventually. Now I can chuckle over plotting and planning tricks on him and he can chuckle over the smashing deeds he did.

I Meet My Husband
(with Bob Steadman)

After I became established at the airport in Flint as an instructor, a new group of interesting people entered my life. Bob Steadman was a young attorney working on a very important murder case. He needed some diversion, so a fellow attorney, a student of mine, recommended that he take flying lessons from me. At first, I couldn't understand what was wrong with my new student. Every time we went up in the air, he would start chuckling. He had been sure he would be too scared to fly. When he found he wasn't frightened, he was just so tickled he couldn't stop letting out the relief.

(Bob S.)I had only been in a light airplane once before I met B. My Uncle Glen worked his way up from being ninth assistant in the great Bethlehem Steel open-hearth mill in Baltimore and was head of the plant when mother hired a pilot to take my older brother and me up to see the size and scope of the thing. I remember being scared absolutely brainless when the pilot banked the plane over the mill so we could see better. Before flying that first time with B, I flew commercially a great deal while in the service, as well as after, with no problems. It always bugged me that I had been afraid as a boy in a small airplane.

The scope of my desire to learn changed dramatically once I saw B. When we went up for that first lesson I really didn't know how I was going to react. With B flying, I realized not only that I wasn't afraid, I was excited and thrilled and pleased. Once I started to laugh, it was so pleasant I was laughing almost uncontrollably. B thought she had a nut on her hands. She wanted to get back down on the ground as fast as she could.

I soon admired B, respected her, and thought she was the greatest thing to enter my life in a long time. So I tried to get her to go out with me. She carefully told me she didn't date students, but I was trying some interesting cases then and I got her hooked on talking with me, so we eventually went for dinner.

I *never* dated students. Well, almost never, but I was interested in the case Bob was handling because it was similar to something from my own life. When he asked, "Would you go out and have a drink with me while I tell you what happened today?" I went. He was able to speak so well, and the case was so fascinating, I was smitten.

However, he almost killed me once. I was propping his airplane. He got a little excited and shoved the throttle in full, which caused the plane to lurch forward toward me. This *really* excited him, so he jammed on the brakes. By then the plane had enough momentum to stand itself right up on the prop! Fortunately, I got off to one side while the prop ate itself down to a three-foot length. My marvelous war surplus leather jacket kept me from being killed by the wood he threw at me.

After that, my friends were full of advice. "My god, you're not going to let him back in that airplane again!" "Forget teaching him how to fly."

"I wouldn't marry him. Go away for a weekend if you have to, but don't marry him."

Bob mounted what was left of the propeller for his office and I decided we were going to get married despite them all. He's been a great stimulant and full of encouragement ever since. I started my business one December and married Bob two years later in February, 1959.

B was doing great things out at the field, with a fine reputation in professional aviation. All of her students were interested in her succeeding. Anything good that happened was reason for a party. I could see she formed few friendships, but they were going to be for life. There was nothing phony about B. When she and I decided we were going to be an item, her friends were insistent in determining whether I was the right person for her. "Watch out. Slow down."

While at Flint Flying Service in the forties, I came up with an idea that worked so well it was also done in a number of communities with great success for a number of years. I know we did it for more than ten years in Flint. I called it "Penny a Pound" Airplane Rides for the March of Dimes. We set up a scale and weighed everyone who came out for a ride over the city. A two hundred pounder paid two dollars for the ride and the kids got a great ride for just pennies. All the money went to the March of Dimes. Our pilots were all commercially rated volunteers who donated their time, the gas and their planes. The local paper, *The Flint Journal*, covered each event with great publicity so that the whole town knew when it was going to happen. Afterward they published how much money we raised.

We had long lines of people waiting for rides. We would start in the morning and fly for hours. I don't know how much was raised overall, but I know the local chapter of the March of Dimes was pleased with our success.

I will always remember how happy the folks were as they participated in the event. It was like a carnival. People who had never flown before came with their children and the weigh-in took on a life of its own. Groups would stand around the scale ready to kid each other as we very ceremoniously called out the weight for the paymaster. If a woman did not want to be weighed so publicly, she could simply pay two dollars to the paymaster for the ride. Many of the children paid for the ride with pennies from their savings. My students helped by controlling the crowd and assisting entry into the planes. I remember particularly how much fun my student, Bob, had conducting the weigh-in. He was an enthusiastic supporter of the March of Dimes because his older brother, Lew, had suffered a terrible battle with polio some years before. "Penny a Pound" was just one of the finer points of life Bob and I discovered that we shared.

Pretty as a picture.

I fell in love with B very quickly. I didn't have any question about what I wanted. I'd been around successful women all of my life, so her success was no concern of mine. It was icing on the cake. We have never had any problems over the mixing of careers. We continue to carefully work very hard at our marriage. I started flying with her after she started her business and helped, when I could, as she expanded it.

After B signed me off, I flew around the state for three or four hundred hours in a little Cessna 170 she sold me. A client in the septic systems business was constantly being sued. I soon came to the conclusion that if you were going to fly on business, you had to spend the time to be professional. It was much easier to hire a professional pilot. And, by then, I had the greatest professional pilot in the world in my household. I also had already learned that, if you weren't her student, when you got into an airplane with B, you didn't fly it!

West, Then East, With a Meeting in the Middle
(by Bob Steadman)

One of the reasons I was attracted to B was that she respected her ancestry. The people in my family were Dakota pioneers. Great-grandfather was an English servant who married the nobleman's daughter, so they were banished from England. Great Grandmother Steadman was reputedly a magnificent horse-

woman. They came through Canada to South Dakota in the 1880s. Their son, the grandfather who I'm named after, became one of the most successful farmers in South Dakota. As happened often in those days, Grandfather's first wife died after bearing three children. Her sister traveled out to take care of those children, married my grandfather, and they had two children of her own. My own father was the last of those two children. Grandfather's holdings were wiped out along with everyone else in the Great Depression with five-cents a bushel corn. When the locusts took over the plains at the age of seventy, he started all over again.

Bob looks great in legal duds, too.

My father was the only one of Grandfather's five children to go to college. Dad went on to get his doctorate in Chicago and has had an extraordinary career. He taught governmental budgeting at the doctoral level and became a leading professor at the Maxwell School of Citizenship at Syracuse. He left there in 1947 to teach at Wayne University in Detroit. In 1948, Dad was tapped by Governor Williams to be the head of the Department of Administration for the State of Michigan. Later he went with the American Management Association, pioneering some of their seminar work and worked for MacNamara at the Defense Department when they were closing defense bases. Dad then wrote a number of treatises for the think tank Committee for Economic Development before he retired.

With five children, Mother was the glue that held the family. She was a brilliant woman. She, like B, wanted to become a doctor, but Dad and she met early in college. There wasn't enough money for both of them to go on, so *she became active in politics and just about anything else you can imagine. Mother and Dad were on a par across the board, sharing everything. Evenings, when he'd come from teaching at Syracuse, we would sit a long time as a family at dinner together. Discussions at the table were about politics and what was going on the world. In those days without television, I had radio, I had my parents, and they were both fantastic influences. As I became a lawyer I realized I had such a break, given a richness of language and thought at an early age.*

I was born in South Dakota, but lived most of my life as a child in central New York in a little town called Marcellus. I finished high school there, then attended Weslyan in Connecticut on scholarship for a year. When Dad took a sabbatical in Washington, D.C. I took a year of study at Syracuse. Then we all came out to Detroit, where I finished my law degree at Wayne University and went into the service. When I was discharged in 1953, I took the opportunity to practice law in Flint and, finally, met the most exciting woman in my life.

Bob came to our marriage equipped with two beagles. Having two dogs in a new apartment complex took a bit of doing. We went over the fence onto some vacant land and built a kennel. Just as he learned I was Pilot-in-command, I learned that dogs are part of Bob's life.

When I came down with hepatitis and had to sit at home, doing nothing, he gave me a golden retriever puppy. That puppy and I became very good friends as we basked in the sun. I pretty much had to turn my business over to a young man when I became sick, and when I tried to hurry back to work, I suffered a relapse. That really laid me up. I felt between the devil and the deep blue sea. That is when I took that cruise with Janey Hart and her gang.

When the astronaut thing was canceled, I didn't know exactly what I wanted to do, but it was becoming clear to me that I was ready to assume a more traditional role in life.

CHILDREN

I don't want to give the impression that children were an unknown for me up to this point. On the contrary, I enjoyed the children in our families. When we used to go to the old airport in Sault Sainte Marie, near my grandfather's farm to pick wild strawberries and watch the DC-2s come in, I never imagined the places I would go and the sights I would see from the air. Later, I tried to share the benefit of my own experiences with the children in our families. Once I took two lively ten-year-old nieces, Elizabeth and Carol, out to the start of a race. On the ferry to explore Vancouver, I let them wander by themselves. Over the radio I suddenly heard, "Will the mother of the ten-year-old girl please come and get her from the engine room." Later a group of us gathered in my hotel room where one of my nieces was reading. In the midst of our conversation, she called out, "Aunt B, what's a prostitute?" My friends thought this was hilarious. I don't remember how I answered her question.

Carol, Aunt B, and Elizabeth.

These two girls are now impressive women. Carol graduated from Brown University and is an investment banker. Elizabeth, a lawyer, represents the Marshall Islands at the United Nations.

Another time I took my ten-year-old nephew Denny to a race with me when he was at the age where every man he saw wearing sunglasses was a crook because "the TV said so". When a couple of fine-looking men wearing dark glasses came past my group at the pool Denny said, "Aunt B, your friends were speaking to crooks!" I was amused later when I put him on the plane and watched them seat a real crook in shackles beside him.

Adventure Remembered Through the Eyes of a Child
by Dennis Whipple

I was ten years old when my Aunt B asked me to accompany her and her co-pilot, Mary Clark, on a cross country flight to California where they were to begin a women's air race. I remember standing on the wing of Aunt B's Piper Comanche at Bishop Airport. It was sunny and warm and I was a "husky" ten year-old kid wearing a pair of shorts and a muscle shirt. Not a lot of muscles under the shirt, though. In retrospect, I can honestly say that I enjoyed flying with Aunt B more than anything else I did when I was a kid. On the

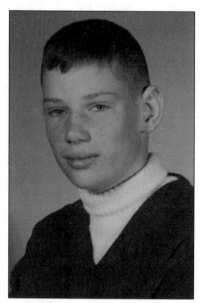

The muscleman in school.

verge of a week-long flying adventure, I suppose I would compare my excitement to the feeling an adult might have as the holder of a winning lotto ticket, driving to Lansing to cash it in.

When the clearance came for take-off and the engine was pushed to full power, as the airplane raced down the runway to become airborne, the exhilaration I felt is even harder to explain. If a person has experienced it, they don't need an explanation. If they have never experienced such a feeling, no amount of explaining can make them appreciate the fine difference in degree of excitement.

Our week together not only entailed many hours in the air, but things on the ground that I'll remember for the rest of life. The layover in Memphis, Tennessee was where we stayed at a Ramada Inn with a swimming pool. I remember wondering if we'd get to see Elvis Presley. I can also remember asking Aunt B if they'd take our American money there. I just about lived in the pool during our stay. It set a precedent that I insisted on everywhere else that we stayed. No pool? No stay. Not only was I husky, I was a brat. Aunt B and Mary Clark had the patience of Job to put up with me all that week. I recall staying in Dallas one night and that we made a stop in Arizona for fuel where a jet fighter was doing touch and goes. I was fascinated to see one up so close. When we flew over the Rockies, I thought I should probably use the oxygen mask. Instead, I slept like a rock and don't remember much about the mountains.

I also remember that for some reason, I was paranoid about crooks on this trip. To me a crook was any guy wearing sunglasses. I felt we were going to be attacked and robbed. Of course, no one bothered us.

In California, I was absolutely fascinated with Disney Land. I'm glad I saw the place as a kid because as an adult, I seem to have lost some of the awe I experienced as a ten-year-old.

When the time came for me to fly back to Detroit, I boarded a Delta *flight at LA International. I was a little apprehensive about flying back to Detroit all alone. I didn't let anyone know that, though. I had an image to protect. Finally I was settled into a section where there were four seats; two seats faced the other two like those in the back of a limo. The guy sitting next to me didn't talk to me at all. He only communicated with the man sitting across from him. They kept passing a paperback book back and forth that was covered by a paper bag. I thought, "dirty book." At one point my seatmate took a dollar bill out and folded it into a fancy design. I thought, "This looks like a skill that could be developed by someone with a lot of time on his hands. Someone in jail, for example." Then I scolded myself, "Crook mentality." Imagine my surprise when we landed in Detroit and the police took the man who had been sitting next to me out of the plane in handcuffs. "A crook!"*

On the commercial flight home, I took a lot of pictures from thirty-five thousand feet. I can't remember exactly what my dad said about the cost when he picked up the pictures from the pharmacy, but I remember his tone and demeanor. Seems like there may have been one or two four-letter words used also. Those pictures were good, too. Lots of undistinguishable land masses and the wing of an airplane.

Denny is my brother's oldest son. I laughed when I called to check that he got home safely and learned that he hadn't noticed the only real crook on the trip until the end of his flight. Denny was also the one who participated in an experiment with me. When he was too small to see over the cowl, I taught him to fly to see if a person could learn safely using only the instruments. It was a success. He became a good pilot.

DOMESTICATION FINALLY ARRIVES

When I was ready for some children of my own, Bob and I began to talk with the Michigan Department of Social Services. From cases he had in family court, we knew there were children who needed loving homes and we felt we had love enough to share. When asked if we would be interested in a special child, we found one child was particularly in need of parents. He was four-and-a-half years old and considered to be unadoptable because he was born with a hole between the valves in his heart. We figured we had to adopt him as soon as we could so he would feel we were his mother and father when he went through the surgery to mend his heart.

Darryl came to us one Mother's Day when we lived in Montrose.

From that point on it was my desire to get the business to where I could sell it, and domesticate myself completely. After I sold the business, we moved to Ann Arbor, following Bob's work, then on to Traverse City. I only kept a TravelAir from my business for a while as my personal plane.

And I soon got hooked on domestication. Darryl was a beautiful child, but just as white as this paper. He had no color because of his illness. He carried a Mother's Day card when he came to us, full of hugs and kisses. I soon found out he was a dynamic child, absolutely going full tilt. He had never been around men before, but he took to Bob right away.

A busy boy and his elephant stick.

Bob had Darryl riding on his lap one day as he mowed my father's lawn. Bob became distracted and Darryl reached for something through the spokes of the wheel just as they rounded a turn. Before Bob could act, Darryl had broken his arm. Darryl had been ours for about a week and there he was, broken already! When I joined my son and his father at the hospital, holy mackerel, I could hear our little boy screaming. I rushed in and couldn't believe what he was saying. "Mommy, Mommy!" My heart sank. I didn't know who he meant. Then he reached out for me: "Mommy, Mommy". As my son calmed, tears just poured down my face. It was one of the great experiences of my life.

After selling my business, I acted occasionally as Bob's behind-the-scenes assistant. Once Bob and another attorney had a case in Watertown, N.Y. We knew we would be gone two or three days and we didn't want to leave Darryl at home. I had been preparing him to fly for weeks by fastening a seat belt around his waist. Still, whenever I thought of flying with him, all I could imagine was that about the time I would get the plane airborne, he would have both engines feathered. He was just that way...zip!...into whatever attracted him, with his mother left to pick up the pieces. But I felt I'd finally gotten him grooved into how to behave in my airplane, so we decided the three of us could take this trip together. If I had been a more practiced mother, perhaps I would have known better than to feel calm.

On the trip to Watertown, our little guy had his arm in a cast. It didn't slow him down much. He was dressed for the flight in a one-piece red knit suit. Once we had him settled in the plane and were focused on the flight, he was so warm he opened the window—in midair. I lost my charts. We finally got the window shut and explained very carefully one just does not do that. After we landed, Bob went in one direction, I was going in another and, Darryl went in a third. The other attorney just stood still and watched us.

Instead of coming with me to get a rental car, Darryl went out to see what was going on around the airport. Before I really knew he was gone, I saw him headed out the door. At the same time, he saw me coming and made a beeline for a glass-enclosed telephone booth. The phone booth door shut. I couldn't make it open from the outside and he was too small to open it from the inside. I was full of the furies at first. Then suddenly I realized I had the little dickens just where I wanted him. "I've got you now!" "Let me out!" "You are right where I want you, you little dickens, now figure this out." Big tears came down his cheeks. Then he finally stopped crying, looked more closely at the door, slipped his hand under it, gave a tug and came out for a hug. We sat on the curb and laughed until tears filled our eyes.

Later, when we went to the hotel, he locked the bathroom door and closed it from the outside. We could not prepare for a three-day trial without this facility. The management had to take the casing off the door.

Another day John Crawford, the other attorney, and I took Darryl shopping while Bob was busy with the client. All of a sudden, my new son was gone again. I couldn't find him anywhere. John and I were both furious. Then we saw him, outside, sitting on the curb again, with his feet in the gutter watching the cars zip past not more than a foot away. I nearly fainted.

A Flying Family

Darryl flew with us as a family in a Michigan SMALL Race. As we formed the crew, we wore sweaters and slacks to match. Our little boy stayed up late the evening before the race, so at five-thirty the next morning, when it was time to get going, he was tired. In all the delay of getting off the ground, he fell asleep in the back seat. We flew the entire race, even the pylon-like turns, and were on the ground for half an hour before he woke up. His only memory of the race is of Julie Auerbach, who was in charge of the event, driving around the airport with him on her lap, letting him steer.

When the FAA administrator came later to Lansing, the women pilots in the Michigan Chapter of the Ninety-Nines created a nice affair. A man who made a science of paper airplanes had just written a book on very sophisticated ones that taught kids aerodynamics. We created a contest, just for the girls, to fold "cross country" or "aerobatic" airplanes, then fly them in a competition. When they decided to let Darryl enter an airplane, he won!

The husband of a Ninety-Nine is known as Forty-Nine-and-a-Half. Beckey Thatcher, an active early member of the Michigan Chapter, set up the children in our chapter as the Twenty-Four-and-Three-Quarters Club. At the end of this contest she presented Darryl with a sweatshirt in Ninety-Nine blue sporting a big new emblem, 24 3/4, and a diploma. At the ceremony, Darryl stood on a table for an interview. There was no one to prompt him as the FAA administrator asked him questions.

"Does your mother fly?" "Yes." "Does your father fly?" "No." Then he looked back at Bob. "You don't, do you, Dad?" "Do you know what makes an airplane fly?" "Yes, I do. The propeller turns and pulls the wing through the air and that's what makes the airplane fly." That ceremony was a big hit, especially in our family.

The last Transcontinental I flew was just before Darryl's heart surgery. Actually, Bob called me in Monroe, Louisiana, when I was enroute. The pressure on Darryl's heart had suddenly gone beyond the limits and they had to operate immediately. I flew directly to Detroit. I lived for a week in my racing clothes, washing them out each night. Once we finally knew he was going to be all right, I went home to Montrose and drove back and forth every day. Darryl walked into that hospital swinging his suitcase, whistling. After his operation, he came out of the hospital, whistling. The only difference was that he didn't carry his suitcase and, he never cried.

While I was the president of the organization, most of our travel was on Ninety-Nine business. After we got to our destination, we always knew most of the Ninety-Nines. But one time we didn't know the any people where we were staying. When we got to the hotel, Darryl was suddenly saying hello to everybody. "Why in the world are you doing that?" With little boy confidence he replied, "You always know everybody."

Nineteen sixty-eight was a busy year. We went to Alaska. Bob was on business for Airway Underwriters, a company in which his legal partner was president and Bob was trial counsel. I went along at the invitation of the Ninety-Nine's Alaska State Chapter because they had never been visited by an International President.

(Bob S.) I think the majority of my most interesting cases from a legal point of view were for that aviation insurance company. I would go around the country pulling together the reasons behind aircraft accidents. I'd try the cases with the help of local counsel to be certain we knew the local laws and customs. Our company wrote policies for small fixed-base operators. We had insured several aircraft that had crashed in Alaska. What made these accidents happen? What should we pay as our share of the damages, if anything? B and I flew north so I could study the landscape for those cases.

While Bob was busy with the accident cases, I was busy with my hostess, my son and her dog. This dog was only one generation removed from a wolf. A wild wolf. In the house of my hostess! She was the person who invited me to Alaska. As a special treat, she served a meal of all the delicacies of the north: salmon they caught and canned, moose meat and bear meat, both home-shot. It was probably delicious. All the time I was eating I was worrying about Darryl so much I don't remember the taste of the food. The dog was snarling. Darryl was laughing. He would not let the dog-wolf alone. I was sick with worry.

We drove to Fairbanks from Anchorage and then rode a train for return. This gave us a good look at the Brooks Range from two viewpoints. We also drove to a glacier. Darryl walked cheerfully and unafraid along the top of it.

There have been few places where I really would choose to live in the world. From that experience, one is Alaska; the other is Traverse City. When Bob's Airway Underwriters-Michigan partner died, the business was moved to Texas and we decided to relocate where we really wanted to live. Traverse City won out.

Located at the base of Grand Traverse Bay, Traverse City has been a wonderful place to live. We have shared almost every kind of experience a family can have here. When Darryl wanted to become a Cub Scout, I became the den mother for a gang of hyperactive boys. These boys were not welcome in other dens because they had superabundent energy. I kept them so busy they won everything we entered. I remember a play they put on that imagined past presidents as astronauts. That was a great success. In appreciation for my work with the Scouts I was inducted, even as a woman, into the Boy Scouts of America. I treasure that membership card.

One evening I went in to wash Darryl's back when he said, "Now mother, I'm going to be all right tonight. You and Dad just go get me a brother. It's all right." "Darryl, do you know where brothers come from?" "Sure. You just go to the social worker and she will get you another one." Thus, our little boy planted a thought, which grew eventually into our adoption of his brother, Michael. It was interesting to me that he remembered Sarah Sarat, the social worker from downstate who had placed Darryl with us. She helped us again by forwarding all the old paperwork up to Traverse City.

A Teenager Arrives
(with Michael Steadman)

Michael came to us when he was about twelve years old, five years older than his new little brother. It was quite an experience to suddenly have an older child in the family. I had only been 'domesticated' for about six years, and Mike wasn't sure he wanted to leave his old hometown. It was near enough to where we lived that we kept getting calls to come and pick him up because he'd run away again.

(Mike) Mom, it wasn't that I ran away. You may have felt that it was, but all I needed was to go visit my buddies. Eventually you and Dad realized this. I had never been away from small-town northern Michigan and suddenly I was traveling all over the place with you as you were trying to sell those round aerodromes. It was the constant hours in automobiles and airplanes and strange cities that made me seek my friends.

Leaving us finally happened so often that we wondered whether it was wise for Michael to remain with us. But then, I remembered how much I felt like a caboose when my mother remarried. I was not going to have history repeat itself with someone I felt responsible for as a member of my own family. Eventually we all worked out the adjustment.

There was the one time...a buddy and I hid out in a barn. We didn't know my best friend had called my dad to let him know that we were fine. The two of us were safely camping out, raiding the kitchen of the nearest friend's house. We thought we'd really pulled one over on them all.

When Darryl caught the wanderlust at age eighteen, I could have creamed Mike, but instead we applied what we learned from working things out with him. When the Circus came to town, Darryl somehow managed to leave with them. But our younger son must have learned from Mike, because he called us from each destination, just to let us know where he was. Darryl likes people, unique kinds of people. He made friends with an old man who made sure he had food and shelter. In all he was gone a month or so, caring for the animals. We were thankful when he got back home.

It is not easy for me to pinpoint when Mike finally began to feel we were his family. A teenager is often not as demonstrative as a hyperactive preschooler is, and Mike was certainly already a more formed personality than our younger son was. Part of what probably made it eventually work was how Darryl adored his very own brother.

He may have adored me right at first, but we came from such different backgrounds, we were pretty unconnected for a long time. It wasn't until recently that we've really begun to communicate as brothers.

Mike's adjustment was also complicated by all the changes in my own life. I won my final international race the year he came to live with us. Mike never flew with me when I was acting as pilot in command at all.

I do remember we began to realize Mike's true capabilities by 1975, when the doctors finally indicated they were going to let me go home after some surgery in Ann Arbor. We had no vehicle with us. Mike had just turned sixteen. Bob called, "You and Darryl get the stationwagon and bring it down. We'll bring your mother home." They had to search the house for the keys. Once found, Darryl, who always thought he could do everything better than anybody no matter what his age was, talked with full court volume all the way to the hospital. This kept Mike awake, but

Darryl and his big brother, Mike.

also made him very tired of his little brother. But that is who Mike is; he decides what he can do and then he does it.

Mike began working in the food service industry as soon as he could. I can hardly boil water, but Mike and his recipes have been in publications at various times ever since he served planked whitefish to the old Paul Bunyan Clan. He's a graduate of the Culinary Institute of America in Hyde Park, N.Y. He's now returned to Traverse City with his wife and two children. It is such fun to have those young people around with all their energy. Mike even worked for a while in construction while he and his wife looked for a business to buy.

We were looking for a campground to run, taking advantage of the beautiful country around our home. After twenty years in food service, it had changed so much I felt almost ready to be done with it. I thought that if I could just be chef, it would be different, but to make real money, I'd have to be in management. Today very few people we work with are truly interested in the food. It is mostly temporary help, intending no future in the art. I was really discouraged until Patty and I found a small restaurant for sale in

Traverse City. What a different world we have now. It is just large enough for us to work as a family, hiring very few employees. We also are developing a respected catering business. Patty is a dessert chef of stature; our son Josh helps when he is in town and not in school. If she's interested, when Melissa gets older she can help also.

Mike is a bit like I was. When I was the sales rep for a round hangar concern, people weren't as excited as I was about the efficiency of the idea. Mike and I have both tried a few options. Now it seems he's found his niche, and I've found a balance in mine.

Chef Michael Steadman.

After a while I could no longer justify keeping even my TravelAir. But by that time I really didn't covet instructing, or even flying, for a career anymore. Then came the operation that knocked me out of the air for good. This was an adventure I'd rather not have experienced.

Surgery

In 1965 I was asked to fly non-stop with air-to-air refueling for a round-the-world flight in a Lear Jet with a member of WACOA, Jacqueline Joo. To prepare for that flight, I went for high-altitude training at Wright-Patterson AFB in Dayton, Ohio. Part of this was in a pressure chamber, to test my physical performance at altitude. It measured my tolerance for flying in the low atmospheres of the jet stream. During these tests, I learned about the effect of flying too high and how to recognize when I was reaching my limit for unaided breathing. Once I was near this point, I performed tasks to see how the edge of my tolerance would affect my ability to continue to be in command of an airplane. During a rapid decompression, the test was how fast I could get the oxygen mask on. I was the only woman going through the program at that time, with a whole bunch of officers in recurrent training. I became their chosen guinea pig, going off oxygen at twenty-seven thousand feet so everybody else in the pressure chamber could watch the process of someone passing out. I never did pass out, but I got right to the point where I would have gone out before they put the oxygen mask back on me. I decided then that if I have to die, oxygen deprivation is the way to go. I remember the far edge of the test as a very pleasant sensation.

Survival training, getting water out of the ground in the desert and such, was also part of the work at Patterson. I was glad to be included in everything, just as if I was one of the officers attending for recurrent training as an Air Force pilot.

BROWN SHOE, BLACK SHOE AGAIN

About this time, we were talking with a fellow in Traverse City who was originally from Flint. He knew a lot of the same people we knew there

Department of the Air Force

CERTIFICATE OF TRAINING

This is to certify that

Bernice T. Steadman

has satisfactorily completed the

Academic Portion, Base Survival Course

Given by

Ground Training Division (EWOT) Wright-Patterson AFB, Ohio

10 November 1965

EVERT L. KNIGHT, Acting Chief
Ground Training Division

I'm a survivor!

and was talking about starting up a women's wear business, to be called "The Kloze Klozet". He wanted us to invest and to form a partnership with him. He would operate the business. Since he appeared to have a substantial amount of experience and the investment wasn't too large, we decided to give it a try. Eventually, after we got pretty involved with the thing, we made the mistake of allowing him to continue to run the business. We finally became concerned about whether we were making any money. One day we made the good, or bad, mistake of going to the store and opening up a desk drawer. It was full of bills that had never been paid. I mean it was a total drawer full. We were both shocked and angry.

To complicate matters, "The Kloze Klozet" had just entered into a long-term lease for a different building on Front Street. We couldn't get out of it. All of a sudden the business was falling apart and facing a huge debt. Unless we developed some income, we would be in a difficult situation.

Bob asked, "Why don't you just take over the new building and run it?" The man from Flint left; I moved in. I sure hadn't learned much about clothing in the air business, but I knew business is business is business: earn more money than is spent.

The building we leased was an old shoe store, with rough-sawn lumber on the walls and a dark atmosphere. Imagine what those walls would do to the fine fabric of women's clothing! I worked with an architect to redesign the inside. When we opened in the new location, we were "B Steadman and Co".

Then I went to market with Margaret Royce, who had worked with "The Kloze Klozet". She was the only one from the original business that still held my confidence. In time, we worked out a fine business relationship.

Our first "buy" for the new business was in Chicago. I was the proverbial neophyte. My understanding of the overall clothing business was that it was filled with generations of talented, well-established Jewish families. I felt if they liked me, they might help. But if I started out again with one brown shoe and one black shoe, I might as well figure out some other way to help raise the money for our still-growing debts. When we arrived at the first display, a man came up right away to ask, "What's your opening?" Margy didn't say a word. Finally, just before the silence became too awkward, I simply told the truth. "It depends on what we see." Fortunately, that was an appropriate reply to a question that asked how much we were going to spend. By the time we left Chicago, they gave us exclusives on some of the best lines in the business. Clothing reps also began to come to Traverse City from Chicago and New York to do clothing shows for the best of our customers.

Although this business was thrust on me, B Steadman and Co was another example of taking a chance on failure but having success beyond my expectations. For someone who feels clothes are mostly for modesty and warmth, that business turned out to be a lot of fun. Eventually I was even honored with a "Business Woman of the Year" award from the local Zonta Club.

Turn Down the Heat

Then, in 1975, came the day at the store when I suddenly felt extremely hot. In our little bathroom, I couldn't get my clothes off fast enough. Finally, I realized what I was feeling would not end by itself, and was pretty serious. After struggling to put my clothes back on, I went out and called Bob. He rushed me to our doctor, who gave me a strong physical treatment, believing that I had a neck problem. It was getting close to Christmas. After resting a day or so, I felt quite a bit better.

The night before Christmas Eve, I had a terrible attack of pain in the neck and head and almost unbearable heat. This time Bob rushed me to the hospital, where our Internist ordered a spinal tap, fearing spinal meningitis. Intense pain. They put me into isolation. The tap showed a considerable amount of blood. Dr. Wunsch shared the results: "Well, the good news is that it isn't meningitis. The bad news is that you are bleeding in your skull, and the situation is extremely serious. In fact, it is life threatening. We either have to get you to Ann Arbor's University Hospital or to Blodgett in Grand Rapids." These two were considered the best for neurological surgery. I was perfectly alert. Bob and I answered together that it would be Ann Arbor. When we lived there, we formed a fine opinion of University Hospital.

The next problem was how to get there. Bob and I assumed flying would be best, but Dr. Wunsch immediately told us flying would not be wise. I knew the airplane they would use as the air ambulance was a twin Beech with a narrow doorway. "She should not be bumped or shaken. Any increase in the bleeding could kill her. Even the trip by ambulance will be dangerous. She must be totally still. There can be no sudden moves or shocks." This is what really made us finally realize how serious my condition was. Looking back, we also realized how very lucky I was to be alive after the strong spinal manipulation by the other doctor.

On Christmas Eve, they wrapped me carefully for the ride south in the ambulance. I was cushioned and strapped to prevent even the slightest movement. Bob got into the ambulance with me for the two hundred and fifty-mile ride to Ann Arbor. We had traveled less than fifty miles when everything electrical on the TC ambulance quit. It was already dark. We were just south of Cadillac, fortunately over to the side of the road.

As I grew colder, with no way to get heat into the ambulance, the nurse continued to voice concern that I was not to be jarred. Immobile, I knew I could die at any minute, but I spoke up. "I know what the matter is. You've got a stuck solenoid out there. If you'd just take your shoe off and hit that thing with the heel, we will get it unstuck and we can go." The driver thought I was delirious, but Bob thought it was a damned good suggestion. Bob continues to laugh about this point to this day. He says I was so full of painkillers I was floating in the clouds.

Finally, they flagged a passing motorist with a CB who called Cadillac for another ambulance. When it arrived, Bob and I assumed we were finally on our way to Ann Arbor. The new driver quickly disabused us of that notion. There was no way they were going to take me to Ann Arbor on Christmas Eve. Cadillac needed their only ambulance for their own community. After some additional cold jostling around, off we went in the new ambulance to Cadillac Memorial Hospital, where I was deposited in a corridor. The only remaining choice was to have the Twin Beech fly in from Traverse City to pick me up. Yet again, I was loaded into an ambulance for the short ride to the small airport. There the air ambulance crew shuffled me in sideways, setting me right on top of a heater. Finally warm and somewhat comfortable, we took off again. Here on top of the heater, the heat could not be adjusted, and I was still immobile. That was one hot flight to Ann Arbor.

Again turned sideways for the trip through the door and down the narrow steps, I was deposited for my fourth ambulance ride. Finally offloaded into University Hospital, I couldn't believe that the first thing they did was pack me in ice!

University Hospital

Dr. Richard Schneider was the head of the neurological department. He packed me in ice to keep me alive so we could wait while he planned the surgery. By this time I was so sedated to prevent any movement, I don't know how many days I was in Ann Arbor before they did the surgery.

To this day Bob is reluctant to discuss what the doctors told him before my operation. I do know he was told to expect blindness and much more, assuming I survived the procedure. My survival chances were not much better than fifty-fifty. I was awake enough to know my condition was bad, and somehow, I was determined to beat their odds.

When they begin monkeying around with the brain, neurologists are never sure exactly what is going to happen, because the surgery creates scar tissue and swelling. This can add symptoms and problems over and above the original condition. Dr. Schneider was straightforward and apologetic as he explained my situation and possible outcomes of the operation. We later realized how fortunate we were to have him as my surgeon. When I arrived, he was scheduled to leave within a few days for a year's sabbatical. He stayed to complete my surgery. Although he was one of the top surgeons in the country, Dr. Schneider later marveled at how successful the surgery was, considering the extensive work that his team did in my head. We became good friends as he took special pride in my full recovery.

During all this discussion, I learned we really don't see with our eyes. Eyes are only a transmitter to the brain. The brain decides what we are seeing. Mine kind of got screwed up. When I came out of surgery, I found myself in a scary world. All of my most cherished abilities, to travel, drive and fend for myself, were compromised. My vision had big holes in the peripheral area on the right side. The medicos kept giving me almost continuous eye tests. My frustration increased as I failed miserably to locate the pins they placed near my eyes to test my peripheral vision.

(Bob S.) B's childhood friend, Berniece, came down to help me keep a level head while I waited at the hospital during and after the surgery. We would spend most of the day in the room with B, but sometimes we would go down to the cafeteria for huge dishes of ice cream and talk it all out, filling our conversation with jokes to relieve the tension.

We couldn't bring ourselves to believe B would not recover, but the realities were sobering and the odds were against full recovery. Even after the surgery, when the doctors were so pleased with how well it had gone, they wouldn't give much assurance that B would lead anything close to a normal life. With two young sons waiting for word of their mother, this was about as bleak a period as anything I have ever gone through.

The path to my surgery really started with a simple accident I had when I was still in my twenties, with one of my first cars. It was a yellow Chevrolet convertible with red leather interior. I always washed my car on my parents' driveway, which was cut into part of a hill, with quite a bit of slope. One time as I finished the car, I gathered up stuff and headed for the house. Suddenly I realized my beautiful car was rolling into the street! As I rushed back to get inside it, I hit my head on the edge of the door.

After putting everything away in the basement, I started back upstairs only to realize I was becoming blind. I knew then that I had done something pretty serious to myself, but nobody was home, so I just lay down. By the time Mother and Dad got home, although I had a terrible headache, I was beginning to see a little better so I didn't go to a doctor.

That blow against the top edge of the car door so damaged the veins above my right ear that, as I got older, they stopped handling the pressure of returning blood to my heart. For a number of years migraine headaches were diagnosed. With the surgery we learned I never had migraines. The damaged vessels were bleeding into my spinal column and disrupting the nerves, giving me headaches. Everything I did during the astronaut testing and the training at Wright-Patterson should have popped that vessel sooner. I am grateful it stayed intact long enough for me to collect those wonderful memories.

Finally allowed to go home, directions were still a strange new phenomena for me. Even the simplest fact had to be totally relearned. Time did not have the same meaning to me as it had before. Eleven o'clock became the time to have lunch. I actually couldn't understand why twelve o'clock had to be lunch hour.

Bob had been staying at a local hotel every night. When he came with me in the ambulance, our sons and station wagon were left back home in Traverse City. Mike had never driven a car outside of town. We needed him to bring the wagon to Ann Arbor. An extensive search for keys by both Mike and Darryl included cutting the straps off an expensive, combination-locked briefcase, a gift from Bob's family. It has become a reminder to treasure for life, even though Bob could never close it again. When the keys were finally found, our boys set off alone for Ann Arbor on snow-covered, slippery roads. Once we finally could hug them, their stories of that trip of over six hours seemed more harrowing than our contemplated trip home.

When we got them calmed down, the family and medicos bundled me up again and made me lie on cushions arranged in the back of our station wagon. By the time we reached Flint, I thought I was never going to get home alive. My now-wonderful brother Dick offered us the night in his home. The next day was no better, as off we went for Traverse City. The surgery was a horrible, horrible, long length of time cut from my life. I have never been so happy to arrive home.

(Berniece V.) I clearly remember how frightened we all were about that surgery. B was such an active person. The accidental bump against the Yellow Peril (named after Amelia Earhart's famous car) had created a nest of neurons, which dripped blood into her spinal column for years. Once her family

got her home she needed more pain pills than she felt she should, so, still in a wheelchair, pilots started flying her downstate to Ann Arbor for pain management treatments. I'd meet her at the airport, wheel her to the car, load her up, drive into town and....get lost, of course!

One time I got on the CB. "We are at the corner of...". A doctor in a nearby car said, "Follow me. I'm headed in the same direction as you are." When we arrived at the clinic, we found out he was a psychiatrist with an office in the same complex. He invited us to make an appointment with him the next time we came down. I can't believe our rude reply: "Thank you, but not if we can help it."

At the hospital, B went in for a biofeedback session. Then we'd pack ourselves back into my car, go to my house in Detroit, lie right down and begin to practice the feedback routine until it was time for her to fly back to Traverse City.

I depended on so many of my friends then, helping with the boys, watching over the house. I had no hair after the operation. My head had been shaved... I mean it was sanded! It took a long time for it to grow back. I could not wear a wig because my head was still too sensitive. The wife of a student from prior years made me a fur hat a little on the large size, and I got through that cold winter comfortably, proudly, with my specially made, wonderful fur hat.

Once I regained mobility, the first time I tried to get to the store by myself, I couldn't figure out even how to get to the alley behind Front Street. I went around the block about five times before recognizing that familiar back door. In addition to directions, numbers were difficult. I would start out to do something, and all of a sudden I would realize that what I was doing just wasn't right. When it became clear I was in for a long recovery, we closed "B Steadman & Company".

Now I was out of aviation, out of flying and out of the clothing business.

Zonta to the Rescue

I had joined Zonta International in Flint after one of the races. This group is a classified business and professional organization like Rotary International. At the time Zonta was formed, Rotary was for men only: Zonta was for women only. Now both groups can be co-educational.

Although a local Zonta Club does not have to accept members from another club, when we moved to Traverse City, Dorothy Helms, the wife of a pilot, arranged for me to become a member in Traverse City.

After my eyesight stabilized from the trauma of the operation, one of my Zonta friends, Ann Mapes, called. "I've got a job for you, if you think you're ready for it." Thus I began working with the government in the Volunteers In Service To America program (VISTA). It paid almost nothing, but it was a job where I had to travel. They paid for the travel time. As the senior energy advisor for Northwest Michigan VISTA, I traveled through ten nearby counties, from the Straits of Mackinaw to Cadillac. The Michigan Department of Energy did the training, so I also had to drive from my office in Traverse City back and forth to Lansing at least once a month. The process of recovering enough to be able to follow driving directions was especially frustrating to me because navigation had been so normal in all of my racing. To force myself to learn, I set up driving challenges for myself. Then finding my way became interesting. For example, I never went south to Lansing the same way twice.

My mileage showed this. To compensate for my irregular route, I figured I'd put in for "x" number of miles and take care of the rest myself. The paymaster said, "No. Put all your miles down. They understand this is a special kind of challenge for you."

With the VISTA job, as frustrating as it was when I knew I ought to know something, I found I could relearn if I was patient with myself and tried hard enough. Sometimes I'd pull over to the side of the road to cry, and I'm not a crying person. Eventually I'd stop crying and kind of plot out the problem. I always got to my destination, and I always arrived on time. The big thing was I learned that I *could* relearn. I could relearn. Then I discovered I even enjoyed learning about energy. I talked to senior groups about energy resources and energy conservation. And, as I traveled around the state, I got my confidence back.

Dealing with money was also a challenge on these trips. Relearning the relationship of seven to eight and eight to nine was a hard experience. I would get to the end of the day absolutely wiped out. Bob had the patience of Job, he really did. We also had a lady who came to keep the house while I was going through all this, and she was a doll, very understanding.

It seemed to take me forever to recover. My job with VISTA lasted about two and a half years until it was not funded any longer. Then I worked in the law business with Bob, keeping the books and managing the schedule.

I'm grateful for the return of my ability. With scar tissue on my brain, I still take medication and will have to do so the rest of my life. The medication takes care of any abnormal neural blips. I don't know whether I could pass a flying physical. I haven't had any visual seizures for about ten years. Fortunately, the medication isn't that strong. Since I'm off all the harsh stuff, I probably could pass flight tests. If I felt I needed to fly, I would take them. But I've found a new and satisfying life.

SCNHEIDER, R. C., B. B. SCOTT, and E. C. CROSBY. *Ruptured arteriovenous anomaly in a former woman astronaut candidate.* Aviat. Space Environ. Med. 50(2):182-186, 1979.

A case is reported of a former woman astronaut candidate who withstood the rigors of the preliminary physical examinations for this position. Some years later, she sustained a subarachnoid hemorrhage from an arteriovenous malformation in the right parieto-occipital area, which was successfully excised. Post-operatively, she had a marked visual deficit, from which she completely recovered within 3 months. The development of psychomotor seizures 5 months later was due to probable scarring in the right parieto-occipital region of the cerebral cortex, the interpretive area for orientation of body image in space, which had been supplied by the clipped right anterior and posterior cerebral arteries feeding the arteriovenous anomaly. These seizures have been well controlled on anticonvulsants. A lesion in the temporoparieto-occipital region due to a hemorrhage from a ruptured arteriovenous anomaly, resulting in the disabling symptoms of disorientation or loss of body image due to impairment of the interpretive cortex, could be devastating to the pilot and a mission. This case raises the question of an automatic use of the CT brain scan in screening potential space candidates, and even the consideration of a percutaneous femoral four-vessel arteriogram in all, or possibly selected, candidates. Most neurosurgeons and neuroradiologists probably would consider the risks of the latter procedure too great to justify its use for fear of permanent complications to the space candidate.

Aviation, Space and Environmental Medicine. Famous where I didn't want to be, but still a contributor in the study of astronauts.

During the period of our client/patient relationship, Dr Schneider was in close touch with my good friend, Dr Stan Mohler, head of the Air and Space Medical School at Wright State University. They developed a friendship over these years. I suspect I was the subject of a number of their conversations. The possibility of my going into space while harboring a weakness in a blood vessel in my brain created the potential for disaster under the uncompromising conditions of space. It was the source of considerable conjecture in the medial aviation-space community. I know both doctors found my success in passing the stringent astronaut physical and the high altitude procedures at Wright Patterson Air Force Base to be remarkable. I agree, since the problems I faced years later could have occurred during either set of tests.

At my last meeting with Dr Schneider, I presented him with a trophy I had won in one of the races for "Best Performance". I had a new inscription made which awarded the trophy for best performance in my surgery. He was moved as much as I was by the extraordinary results of his work and what that meant to me in my future. We parted, both somewhat misty-eyed. Some years later he called me from Chicago to tell me his cancer had spread. I was depressed for a long time over the prospect of his passing. It didn't seem fair that we would lose him after he had saved me. His wife told me that his little trophy had been displayed in the center of his desk as a prized memento.

IWASM

The International Women's Air and Space Museum has been a major part of my life for more than thirty years. The seeds of thought for the museum were planted by Page Shamburger in conversations we had at an executive board meeting of the Ninety-Nines, Inc. The continuing effort to develop the museum has provided an opportunity for me to meet and enjoy many tremendous people.

In writing about the ebb and flow of the project I find myself thinking about Page and relishing the chance I had to become one of her friends. She was an extraordinary person, warm, caring and dedicated to perfection. In her presence no one in the project would ever believe that mediocrity would be acceptable. Together, we dreamed of a museum that would be interactive and computer based at a time when most museums were static exhibits.

Page Shamburger.

After lengthy discussions with members of the Ninety-Nines and with museum professionals, my husband Bob drafted a trust document for submission to the membership at their annual meeting. In 1971, the proposal was simple: "The Trust shall have as its major role, the creation and the operation of a Museum dedicated to the exploits and contributions of Women in Aviation and Space." To ensure the informed cooperation of the Ninety-Nines, the Trust designated the vice-president of the Ninety-Nines, Inc. to be liaison with the Trust, taking part in all deliberations of the trustees. To further emphasize the strong link, the Trust provided that amendments to the Trust could be made by a simple majority vote of the membership at their annual meetings.

The resolution establishing the Museum Trust was adopted at that annual meeting in 1971 while I was President, making the International Women's Air and Space Museum (IWASM) a reality. I remember the moment my gavel fell to mark the occasion with great pride and remain convinced that our work to collect, preserve and inform the public about women in aviation will always be important.

We had to find outstanding people to step in and help the museum organize. I knew we needed more than women who were just famous as pilots. We needed dedicated women with names in aviation who could contribute both their acumen and time, women who would dig in and help us convince other pilots to save their stories. They had to be willing to become mentors. I was pleased by my success. Eight women shaped a strong nucleus for the Trust: Jacqueline Cochran, Betty Gillies, Jimmie Kolp, Blanche Noyes, Page Shamburger, Nancy Hopkins Tier, Virginia Thompson and myself. Betty, Blanche and Nancy were charter members of the Ninety-Nines. Betty served as a WASP in WWII and showed her dedication to the organization as chairman of the AWTAR for many years. Jimmie had a great reputation for business sense, was a commercial pilot and served as Treasurer of the Amelia Earhart Fund. Blanche's association with the FAA was invaluable. Nancy served as an officer in the Civil Air Patrol and had more than fifty years of experience in aviation. Virginia was well regarded as an active leader and pilot.

Olive Ann Beech, Betty McNabb and Doris Renninger joined Jackie, Page, Jimmie and Blanche on the Accessions Committee. An Advisory Committee of men was equally strong with Lewis Casey (Curator, Aircraft; Smithsonian Air and Space Museum), Dean Krakel (Managing Director; National Cowboy Hall of Fame), Grover Loening (aircraft consultant) and my husband, Bob, as legal counsel.

About this same time Elizabeth Sewell presented a concept for a Ninety-Nines, Inc. headquarters building at the Oklahoma City Airport.

To my chagrin, that proposal also gained strength and caused a division of fund-raising effort. This eventually led to a split between the two groups, as the Ninety-Nines voted to concentrate their efforts on the headquarters building.

Progress is built partially on the confidence of known history. Educating people about saving their own history is still the key to IWASM. At first we tried to get other museums to make women a more important part of their displays. It was a great idea except, for a long time, other museums did not want anything to do with womens' part in aviation history. Page became a registered aviation historian. She could go almost anywhere to do research work. She had acquaintances and credentials as an aviation writer, so she really was the guiding light in terms of getting ideas into motion. I was glad to help spread her light.

IWASM logo.

On February 9, 1973, IWASM trustees and officers met at the Smithsonian in Washington with Lew Casey. We were to determine the physical requirements and approximate cost of establishing the museum. Using his vast experience, we developed parameters for the project. We believed we could build the museum in stages.

I went out to ask Jackie Cochran how we should set up the criteria for the location of the museum. She was such a comfortable hostess! She made me feel special and at home. Jackie said the location of the museum should be in a community that demonstrated that they wanted us. Their town would reap benefits from our museum, so they should feel that they have a vested interest from the beginning and should play a significant part all through the development: put up money, give us a building. With her advice in mind, we developed our material and sent it around the country: Oklahoma City wanted us; Dayton, Ohio, wanted us, and Long Island, New York, wanted us.

CENTERVILLE, OHIO

We hired an architect to conduct a feasibility study to determine what would be the best geographic location. The study measured not just where aviation people were, but where the best vehicular traffic would allow us to have the potential to support ourselves. Doris Scott was a businesswoman in the Dayton area. She talked to the Chamber of Commerce. They were very interested. Virginia Kettering, a pilot and wife of the man who established the foundation to start the Air Force Museum, was also in Dayton. With all of this interest, it was clear Dayton met our criteria for traffic and potential.

The architect who did the feasibility study was a student of Alden B. Dow. He recommended Dow to us and introduced us, for Dow had designed a building for interactive display. Alden Dow was a member of the Dow family of Midland, Michigan, the home base for Dow Chemical. In his own right he was a world renowned architect who had studied under Frank Lloyd Wright. His home in Midland is considered to be one of the most beautiful homes in America, along with Wright's home "Falling Water".

I found Alden Dow and his wife both to be warmly gracious and giving. Dow's personal interest in the IWASM project was an important affirmation of its potential worth and stature as an international museum. The committee met in Midland with this man for two days in 1975 to review our ideas for a dynamically displayed museum featuring an Omnimax theatre. Mr. Dow made it clear to us that our desire to build a great museum while having no background in such construction was no impediment. *Reflections*, his own book, which he gave to me, is prefaced with the exact description of his approach to the IWASM board. He wrote, "Every man has the privilege of exploring his own mind for the answers to any subject he may find of interest to him. It matters not whether this idea is his own field. His amateur standing should not hinder him. In fact, it should encourage him. As an amateur he is probably going to find a different approach to whatever the subject may be and he should feel free and be free to express his thoughts." While working with him we were never given the impression that we were amateurs. We were treated as valued individuals with a single focus on our dream. He believed our museum not only capable of being designed and built, but very worthwhile. Mr. Dow was so intrigued with our prospects that he initiated discussions with Disney about creative displays with interactive, sensory experiences. Walt Disney was already known to us as a supporter of women in aviation through his design of "Fifinella", the symbol adopted by the WASPS of WWII. This great man shared the dream of a world-class auditorium built for commercial aviation display. I honestly don't know which one of us was most excited about the project.

Shortly after our meeting with Dow, Dayton indicated they would be willing to give us a site on the Cox Municipal Airport in Vandalia, Ohio. We were all set to go ahead with the throttle in a fully open position. Unfortunately as our project became even stronger and more exciting, its scope became a source of concern for the Ninety-Nines. In November, 1975, the board of their organization indicated to me that they wanted us to drop their name. I was shocked and disheartened, but the trustees of IWASM were ready to continue as a separate organization.

A month later I required the neurological surgery. The doctors indicated I would need at least two years to recover. I thought Doris Scott was the logical person to carry on for IWASM. We still all believed Cox Airport would be our destination. Doris was empowered to form a local committee to begin a fund-raising drive. Sadly the next few years were bad for our national economy and fundraising was severely curtailed. To my dismay, our proposed lease on the airport site was allowed to lapse and the Ohio Corporation formed by Doris canceled the agreements with Alden Dow.

If Dow had been able to continue as our architect, I feel we surely would have succeeded sooner to build a major showpiece. But this was one decision I had no opportunity to influence. Our loss of his services was a terrible blow. When I learned about it I was at as low a point in my adult life as I can remember. Doris put our collection into storage and IWASM was nearly dormant for several years. A few years after my surgery, I was asked to lend support for an effort to put the pieces back together. Nancy Tier and I, with several new faces among the old, went to work. We still believed in the Dayton area as the ultimate site for IWASM.

Susan Schulhoff Lau, another Daytonian, contacted the newspaper. The paper published a great article. The mayor of nearby Centerville, Shirley Heintz, got in touch with Susan. "I've got a place, if it's suitable for you, rent free." So we settled IWASM a few miles south of Dayton.

Asahel Wright, a great-uncle of the Wright brothers, originally owned a building in Centerville. He used it to manufacture hard candy flavorings, called spirits. Small, it is a prairie building made of thick walls of native limestone and is located in their historic district. Doris brought the memorabilia we had in storage to Centerville, and we were suddenly back in business.

The official opening of IWASM was in 1986. We began to build membership from around the country. We didn't charge admission to the museum, but had a donation box. We put on three or four lectures every year and never charged for them. Monetary donations and membership dues were invested. We operated the museum and performed our educational service with the return from that investment. Expanding our collections, we accepted scrapbooks, authentic papers documenting certain events, some clothing, and a lot of books. There was no place to accept or display airplanes.

Nancy became the president of IWASM. She was fun to work with, full of stories. She was only nineteen when she helped found the Ninety-Nines, with Amelia Earhart, Fay Gillis Wells and others. Nancy also was enamored of cars at an early age. Once she drove by herself from the East Coast to California in order to help her mother and her sister. For protection she dressed as a man and carried a German Luger. In those days, the transcontinental highway was little more than a dirt track, and it was not unusual to have to change two tires every hundred miles, the driver patching the spare herself. To look at Nancy in her IWASM guise, you would never imagine her doing anything far beyond the social norm. That contrast created the strength we needed to pull the museum together.

IWASM GOES TO GREECE

In 1987, I went with Nancy as representatives of IWASM to the Federation Des Pilotes Europeennes (European Women Pilots' Federation) meeting in Athens, Greece. IWASM paid for a portion of our tickets and we paid the rest of our costs. We were gone two weeks, including a Mediterranean cruise. Marie Josephe de Beauregard was the founder and first president of the organization. She wrote a history of her countrywomen as pilots and compiled a collection of stories from other countries in *The Pink Line*. In 1995, she and Mary Claire Pelé came to visit us at IWASM in Centerville. She continues to serve as our European consultant, sending us historical information now and then, and fielding membership questions.

When Nancy and I were in Greece, we met Fiorenza de Bernardi, the daughter of the Italian aircraft designer. She was the only woman then type-rated in a DC-4. Marie Fede Caproni, who founded Italy's first aeronautical museum, was also there. I met Ann Sperry, the flying doctor. Her book, *The Sugar Bird Lady*, is a good history about the polio vaccine. I already knew Sheila Scott, who wrote *Barefoot in the Sky*, but Elizabeth Overberry was new to me. Some of the pilots attending the meeting were "Atta Girls" from WWII. From France I'd hoped to meet Jacqueline Auriol,

Nancy celebrates her birthday.

who was constantly trading aviation firsts with my friend Jacqueline Cochran. Her father was an aircraft designer. She had a very bad accident while flying a prototype jet, and it took a long time for her to recover. The story of her life, *I Live to Fly*, is quite something to read. An English version is available if you hunt the used bookstores. We met Eftbynia Bellou who developed a pedacopter, the Polyhelices. She was a taxi driver in Paris, trying to make enough money to continue with her pedacopter, and was an interesting woman. Mary Claire Pele also entertained us later in Paris.

During the meeting of the Federation, I learned there is a tremendous amount of nationalism in aviation. The French think they did it all, the British think they did it all, and we certainly think we did it all. Everybody was happy with aviation and their country's part in it. Aside from that, the trip was normal. Even in Europe I can walk into any room full of people I don't know, and if there is another pilot in the room, we get together. Then you might as well stop trying to entertain me with anything else.

All of the pilots at the meeting in Greece were aspiring to have women pilots accepted as professionals in their own country. Over here we may have been slightly ahead of Europe, but we were all struggling.

On both sides of the Atlantic it was still the privileged few who were getting anything like commercial slots, so all the pilots were trying to get the airline business more open to women, while each was scrambling for her own place as a professional.

We had flown into Frankfurt, Germany, then down to Athens for the meeting. Somehow in the two weeks we were in Europe, we cruised the Aegean Sea, traveled by rail to Paris and returned to Frankfurt to visit Nancy's niece, married to a German fellow, Hermann Hoffmann. We stayed with his family in Bad Dirkhaim, a hot springs near Heidelberg, for several days. Hoffmann's family had an upscale shirt-making business. His family had sent him away during WWII so he could return after the war to head the company. The Hoffmann shirts were good quality.

This trip is just one small part of the fun I've experienced with women pilots from around the world and it helped validate both my work and IWASM.

Dayton, Ohio, Again

By 1995 the Asahel Wright House was getting too small to hold us. We began to look overstuffed, and it was too difficult to locate material efficiently in our files. There was not enough room to display most of the original donated memorabilia. Even the library in the annex was becoming too crowded. IWASM again approached Dayton to ask for larger facilities near the confluence of their four rivers. Some of the buildings in the Wright brothers' old neighborhood were becoming unrepairable, and drug dealers appeared to be moving in. Dayton had begun to seek funding from various governmental sources to renovate the downtown. I think we became the first non-city organized effort to do something. We felt that what we were doing was good for the city and the neighborhood. Of all the times that people need to have history to learn from, they need it most when they are feeling discouraged. We felt we could create a large focal point for the history of aviation that already existed in Dayton.

IWASM began to be projected as the cornerstone for a development adopted by the National Park Service. Two blocks over from our location in Dayton's Aviation Heritage Park was the Wright/Dunbar District, where the Wright brothers grew up. Dayton hopes to restore this area like Greenfield Village in Dearborn, Michigan, or like Williamsburg in Virginia. The Aviation Trail will include Carillon Park where the Wright brothers' original airplane is and the Wrights' bicycle shop. The original parachute factory would have been right across from our museum. They say the Air Force Museum in Dayton attracts more than a million people a year. That museum is on the Trail. We felt that even if IWASM only attracted a fair percentage of those people into downtown Dayton, we would be able to support a larger facility and staff, and retire a standard mortgage.

Gary Snyder, local architect, and his wife Carol, an interior designer, began to work with us. IWASM didn't want to go into its capital to hire them, but they were what we were looking for, so Nancy found the money herself. Gary drew up plans for a very nice building, practical, about the right size and expandable in two directions. A rendering was done for the site. If we moved to Dayton, the building itself was estimated to cost three and a half million dollars. We wanted to have another million and a half for an endowment investment. Then I became president of IWASM. On the top of the building I helped plan Nancy's Tea Room, with the roof used for special-events dining.

The city paid for a site review, which triggered a lot of interest from other organizations and reduced the size of the property available to IWASM. The land we were promised shrank as Dayton located other buildings around our design site. After spending a considerable amount of time with Dayton's officials, it became obvious there was no longer enough land to fully develop our building plans.

Connections

In 1987, IWASM developed an exhibit dedicated to the Mercury Women. The five of us who were able to attend were thrilled when a letter from our country's first woman astronaut was read aloud. Sally Ride wrote, "Although you were not able to realize your collective goal, your accomplishments demonstrated that women could perform and achieve in space and opened the door for those of us who followed."

Another example of how one thing leads to another is the story of a woman I have been known to fondly call "daughter". Amy Carmein is a young woman who took her experience in high school journalism and turned it into a successful business as the owner and publisher of the magazine, *Women in Aviation*, to serve every woman who did anything in aviation; mechanics, gas jockeys, artists, writers, private pilots, commercial pilots, designers. Contributors to the magazine come from a background in control towers, the FAA, jets, balloons, gliders, history, education and adventure.

Several years after the formation of *Women in Aviation, The Publication*, Dr. Peggy Baty, Academic Dean and Associate Vice President of Parks College, spearheaded a Women in Aviation (WIA) Conference. The first conference in 1990 was attended by one hundred and fifty people. I use the term people because men were paying more attention to women in aviation by this time. I attended the conference in 1993, and was a speaker at the 1997 conference, attended by over two thousand aviation enthusiasts, when the women who had passed the Mercury Physical were inducted into the WIA hall of fame. During this conference I heard a pilot comment on riding in the elevator with some women who were in town for a farm convention. The farmers could not believe how many women were staying at the hotel for the WIA conference and they expected the majority of the two thousand pilots at the conference to be men! A few years ago, the international organization which grew from the work of Amy, Dr Peggy and others became *Women in Aviation International, Inc.*, and purchased Amy's magazine, renaming it *Aviation for Women*, with Dr Peggy as publisher.

Both Amy and Peggy worked for IWASM professionally for a while. Beginning in the '80s Amy turned our quarterly newsletter into a professional publication and when IWASM needed some new direction, we hired Dr Peggy for one year to help us begin a fund-raising campaign for a new building in Dayton. She also was in charge of developing a computer base for IWASM before she moved *WIA, Inc* to Embry-Riddle Aeronautical University's Daytona campus in Florida.

CLEVELAND, OHIO

Another option was developed to move our museum to the Burke Lakefront Airport in Cleveland, where the Bendix and Thompson Trophy races terminated. We have always recognized the advantages of being on an active airport. If we located in the terminal building we would not have to raise funding for a new building right away. With the lack of space and the apparent long timeline for Dayton's Aviation Heritage Park, the Cleveland option was the route we chose to follow.

My friend and racing buddy, Joan Hrubec, became the Museum's Director. Once the decision to move was made, it was amazing how fast Joan and the rest of her support group got things up and running.

Connie Luhta became president. A pilot, she owns a successful FBO and is a mover and a shaker. IWASM has hired as many people as possible from the Cleveland area, including a professional display group. Now we also have a full program re-established. We bring people into Cleveland, we're an asset for them, and the city is responding beautifully.

At this point I'm retired as president of IWASM but remain active as honorary chairman and trustee. My main concern for the museum is to create an easy access to the increased displays and to our library of clippings. It will be wonderful to have all the files entered into a computer for research purposes.

And I still have ideas. I'll grant you they are not new, but static museums collect dust. The Air Force Museum and the Smithsonian...well, it is nice to go in and see their displays, but somehow we need to tell a more complete story. When we first started talking about our museum, we went to Silver Hill, the Smithsonian storage facility, and saw Amelia's Vega sitting outside, deteriorating almost to the point of not being recoverable. With IWASM interest, it finally became a major exhibit for the Smithsonian. Still, I would like to see Amelia's Vega in our museum, in a display where you can smell the gas and oil, the leather, you can smell the entire unique mix of elements that make it *her* airplane, you can hear the sounds of it, and see the wild ocean around you. I know museum people can even recreate fire around the cowling. All of these things are what will really make the story ring true. If you could just hear her voice tell part of her own story from our recordings...she had a distinctive voice. Perhaps we can create something like this in Cleveland.

In her notes from a November 3, 1999 article for the *Centerville Times*, Sarah Rickman writes about the history of IWASM in the Asahel Wright Visitor and Community Center: "...when I walked inside last Friday, ghosts spoke to me...the women pilots whose presence, for 13 years, graced the building near Main and Franklin. ...Amelia's picture hung at the bottom of the stairs [that lead to the offices]. Her penetrating eyes took the measure of every soul whose footsteps trod those stairs in search of the history of the women of flight.

"Nancy Hopkins Tier, IWASM president, put her heart and her final ten years of life into making [our] museum a viable reality. ...Nancy and fellow IWASM board member [Fay Gillis Wells] helped Amelia and 96 other women found the Ninety-Nines...Though Nancy is gone now...Fay still works to further the goals of the museum.

"The alive and well spirits...spoke to me as well; Joan Hrubec, B Steadman, Nadine Nagle, Margie McDonald, Astronaut Cady Coleman, Marcia Greenham, Judith Wehn—all boosters of women in aviation."

Sarah did a good job of preserving a glimpse of the museum in Centerville. She could not include everything, but the article gives a taste of what we are about.

Looking Into the Future

I talk with many students now. Some haven't even heard of Amelia Earhart. We were of an era where she was the most publicized woman in the world. For us to live three generations more and they haven't heard of her? We haven't done a good enough job. We have to make her story available to others, especially to school groups, as part of the story of other women in aviation.

Look at what is coming out of the Bessie Coleman story. We need to recover and preserve information to create a strong future. I see people like Jackie Parker test flying jets at Wright-Patterson, Eileen Collins in the space program, Kathy Gotsch piloting the *Stargazer* transport and I think, golly, they're frontiersmen. But you don't hear enough about them, or what made them reach for aviation and space. We need to save and tell the stories of active women *now*. If we don't begin to collect history as it is being made, and focus on the bits each does to advance aviation today, these stories can fade just as easily as they did in the past, simply because there are still fewer women than men in aviation.

My lifetime achievements are an accumulation of many small steps. I think the museum is the same. The stories that need to be told are incentive enough for me to keep working on them. Women have been so under-estimated and under-promoted that there's an awful lot of brain-power and ability that is not being properly applied. People who find they have a gift really have a responsibility to do something with it. I think the stories of today's women are tremendous. Men and women, girls and boys can receive a live coal from the fire of the stories we display at the museum and it will help them blow their own talented coals into a flame. Just keep watching us!

Looking Back to Shape the Future

Bernice Steadman

After WWII both the Army and the Navy wanted to control the new space program. Each began developing rockets and hardware to prove they were the most capable of taking overall leadership. Brigadier General Don C. Flickinger, a flight surgeon, was active in the development of the Army's program. Jerrie Cobb and Dr. Lovelace met with him at the beginning of the clinical testing of the Mercury Women. It is Jerrie's impression that General Flickinger would have been receptive to the inclusion of women in the selection process for Mercury Astronauts. It was clear he shared Dr. Lovelace's interest in testing women for selection.

However, NASA was established as a civilian agency on October 1, 1958. In establishing NASA, President Eisenhower announced that our national space policy was too important to leave to the military or to scientists alone. He commented that, "We must free ourselves of the attachment to service systems of an era that is no more." Eisenhower then appointed Admiral T. Keith Glennan, retired, who was also past chairman of the Atomic Energy Commission, to run the new agency. With NASA established as a civilian agency, General Flickinger apparently was not in a position to intercede for the Mercury Women. Despite the fact that NASA was now a civilian agency, in the final analysis another military man (war and space hero, Marine Colonel John Glenn), had a vital role in preventing our participation in the astronaut selection process.

After so many speeches about the Mercury Women, I am struck by the regularity with which two questions are asked: "Why didn't NASA go forward with a Woman in Space program?", and, "Who stopped it?" Such questions led to further research. It is quite clear that NASA wanted only military test pilots as candidates. Since women were excluded from the military test programs, it was NASA's position that there were no women qualified to compete to become astronauts. NASA ignored the existence of the Mercury Women for as long as it could, then continued its policy of excluding women pilots as astronaut candidates to pilot space vehicles for more than thirty years.

When the story of our testing at Albuquerque finally broke in the June 28th, 1962 issue of *Life* magazine, we heard all kinds of little jokes made about women in space. As these were attributed to the astronauts, we began to realize that the Mercury Seven Astronauts had no interest in competing with women for flights in the Mercury capsules. I really don't know whether their lack of willingness to compete was based on ego or other factors. For example, we read reports that the seven astronauts had dealings with *Life* magazine. It was widely understood they were paid millions of dollars for exclusive rights to their life stories. Since officers in the military services are not allowed to accept private money, they must have been given an exception because of their status as astronauts. I want to make it clear that none of the Mercury Women were motivated by the thought of making money from the program. I have never believed any of the Mercury Seven were either. However, I do believe that the inclusion of a few women in the Mercury Program might have had a negative impact if the reports of financial dealings were correct. The possibility that the first woman in space might be a Mercury astronaut would have been big enough news to overshadow the outstanding accomplishments of the original seven men.

Looking back to the conclusion of our physicals, NASA's decision to cancel the testing and training at Pensacola must have been strongly influenced by its surprise at how many of the Mercury Women had passed the astronaut's rigorous physical. The shock of our success would have been compounded by Jerrie Cobb's successful demonstration of her ability to fly jet aircraft at Pensacola. It must have been a revelation that women could compete if given a level playing field. With NASA's failure to proceed, there can be no question that this was an unwelcome revelation. When we consider that it was another thirty years or so before NASA allowed a woman to pilot the space shuttle, we can begin to understand the consternation our results must have caused. I am unaware of any serious consideration of including women in the Mercury Program until Dr.

Lovelace's report to NASA on our success. It is important to remember the Mercury Women were told by Jackie Cochran and Dr. Lovelace to make no public statements until our program was finished. We were not even to discuss it among ourselves. Remember also that NASA wasn't acknowledging the existence of our testing program. With Dr. Lovelace's conclusions that woman were physically capable of competing, NASA was suddenly faced with a problem it had not asked for, and clearly wanted no part of. NASA's solution to the problem was to ignore us until they were brought before the Congressional Subcommittee on the Selection of Astronauts in July, 1962.

The abrupt end of our program was a great shock. When Jerri (Sloan) Truhill and I met with Dr. Lovelace at the conclusion of our testing in 1961, he told us that we had passed the physical with flying colors. He also made it clear that he was enormously pleased with how well all of the women were doing on the tests. He could not hide his elation because our success fully supported and validated his decision to test women for participation in the space program. He assured us that, if the others continued to do as well as he expected, we would all be going to Pensacola for the additional physiological testing and the jet training Jerrie Cobb had successfully completed. He told us that after our jet training was finished, he was positive that we would be included by NASA in what he described as a Woman in Space Program. His enthusiasm was contagious.

When Dr. Lovelace spoke about a Woman in Space Program we had a reasonable expectation that he knew what he was talking about. We knew that NASA had selected Dr. Lovelace and his clinic to develop the testing protocol for the Mercury Astronauts and that Dr. Lovelace and his staff had performed all testing of the original group of Mercury Astronauts for NASA. It was certainly no secret to us that we were being given the same grueling tests given the men in that program. Dr. Lovelace was also chairman of NASA's Special Advisory Committee on Life Sciences at the time of our testing and later became senior consultant to NASA's Office of Manned Flight and Director of NASA's Space Medicine program according to *The Lovelace Medical Center*, by Jake W. Spidle, Jr.

After our meeting with Dr. Lovelace in 1961, Jerri and I went home from Albuquerque walking on air. He assured us that Pensacola would be our next destination. You can imagine our chagrin and the depth of our depression when the Navy canceled the jet training in Pensacola. I learned much later that NASA had refused to fund further testing or any training of our group.

Then, on February 20, 1962, John Glenn rode an Atlas missile into space in his Mercury Capsule, *Friendship Seven*, and accomplished the first orbit of the earth by an American. Glenn's name for his capsule reminds me of the name of the Lockheed Vega, *Friendship*, that carried Amelia Earhart to her first taste of fame as the first woman to cross the Atlantic in an airplane. The coincidences continue, since Glenn was essentially a passive passenger in a capsule without aircraft flight controls and Amelia was a passive passenger rather than the pilot of Friendship.

When he returned to earth, John Glenn was a national hero on a level comparable with the national adulation for Charles Lindbergh and Amelia Earhart when they made firsts by flying across the Atlantic. His appropriate perks included a trip to Washington on Air Force One with President Kennedy and an address to a joint session of Congress.

About the time of his flight, rumors were filtering through that Russia intended to send a woman into space. During the cold war, the political and propaganda significance of the United States sending the first woman into space would be tremendously valuable. Russia had already won every major first in the space race, i.e. first satellite in orbit, first animals in space, first man in space, first man in orbit, first two-man space flight and first orbiting of two manned space crafts simultaneously. We knew, with our extensive hours as pilots in command and proof of our physical ability to compete, that America's Mercury Women were qualified to become astronaut-candidates. With one of us, the U.S. could win the race to put the first woman into space.

The Hearing

Jane Briggs Hart (my friend Janey) and Jerrie Cobb attended the hearing that was held before the Congressional Subcommittee on the Selection of Astronauts to determine whether such an objective and program should be adopted by NASA. John Glenn, only five months after his triumphal flight, testified at this hearing as the chief spokesman for NASA. As America's pre-eminent hero in space, he, more than any other individual, represented our country's aspirations and pride in the tremendous accomplishments of NASA. He did more than represent NASA. He testified clearly from his personal beliefs regarding the abilities of women as pilots.

Soon after the hearing, Janey told me about Glenn's testimony and that Jackie Cochran had failed to support us. However, the ferocity of the attack on a possible Woman in Space Program was not clear to me until the actual testimony was quoted extensively from the Congressional Record in the recently published and beautifully written book, *Amelia Earhart's Daughters*, by Leslie Haynsworth and David Toomey. The sum of his demeaning testimony was that women, and specifically we women who had passed the physicals, couldn't handle the job of astronaut. He was America's living space icon and none of the congressmen appeared willing to challenge his personal motives, beliefs or the absurdity of much of his testimony. John Glenn's testimony, as chief spokesman for NASA, and as the premiere hero of the day, was the killing blow to the dreams of the Mercury Women.

John Glenn's Testimony

According to the *Congressional Record*, during the hearing while Glenn was testifying, Congressman Fulton said, *"You must remember that Ham made a successful trip, too . . . .I think a woman could do better than Ham."* (Ham was the female chimpanzee who had been a passenger in one of the first Mercury capsules.)

Unfortunately for our dreams of inclusion in the space program, Glenn's response to Congressman Fulton highlighted his personal notions regarding what a woman can and can't do (or should and should not do). It also illustrated the humor the male astronauts shared over the issue of women in space. He answered, *"That is not a fair comparison, sir, with all due respect. I would like to point out too, that, with all due respect to the women that you mentioned in all of these historic events, where they performed so fine, they rose to the occasion and demonstrated that at the time they had better qualifications than the men around them, and if we can find any women that demonstrate that they have better qualifications for going into a program than we have going into that program, we would welcome them . . . with open arms.'* This was greeted with a roar of laughter from the crowd. *Glenn smiled and continued, 'For the purposes of my going home this afternoon, I hope that will be stricken from the record.'* Again there was general laughter."

On the question of why so few women chose the field of test piloting Glenn's response was equally ridiculous. He called the low numbers *"a fact of our social order."* It will come as no surprise that Glenn failed to mention that women were excluded from the military test pilot and jet flight programs in which the Mercury Seven had obtained their experience.

When asked about relative experience and the need for hours of test or jet aircraft experience, Glenn was even less complimentary to the Mercury Women as pilots, if that is possible. He expressed his contempt for our flight experience by suggesting NASA would have to reduce its qualifications to "a lower level" to allow the Mercury Women to compete.

"To say that a person can float around in light planes or transports for-I don't care how many thousands of hours you name-and run into the same type of emergencies that he is asked to cope with in just a normal six-month or one-year tour in test flying is not being realistic. Instead of trying to reduce our qualifications to a lower level . . . perhaps we should be upping the qualifications and saying that we have to have test pilots with doctorate degrees and with even more experience than we have had to date so far."

His reference to upping the qualifications is very ironic because NASA had waived its requirement for an engineering degree for both Glenn and Carpenter on the basis of equating their test pilot experience with the required diploma. I have always believed that the waivers given so easily to them could have been offered to us, given the extensive flight experience we all had.

It is clear that Glenn made no real attempt to find out anything about the Mercury Women and had no idea what they really had accomplished in their careers as pilots. He also appears not to have had a clue as to what it takes to fly private aircraft as an instructor, as a charter pilot or as pilot-in-command in a commercial capacity. Even more obvious was his memory block on the accomplishments of the WAFS and WASPs in WWII during a period of time when he would have been old enough to know firsthand about them. Most striking, in the context of the questions, was his failure to note Jerrie Cobb's success at Pensacola where she decisively demonstrated her ability to fly the jets.

Jerrie Cobb, when asked whether jet training would have been necessary for the Mercury Program, has often pointed out that the Mercury Capsules were not configured anything like a jet fighter. They had no jet engine, no wings, and no flight control surfaces. Given the minimal controls included in the capsules for intervention by the astronaut, Glenn's testimony underscored the fact that NASA's reliance upon the jet pilotage requirement as the non-qualifier for women was essentially bogus.

In 1962, all he did was take a marvelous ride. He was strapped into a capsule and launched. The story of Glenn's ride is documented in *We Seven by the Mercury Astronauts Themselves*. Although no one can doubt his personal courage, there is nothing in his background as a test pilot that was used to influence the success of that ride. Although Glenn had flown in two wars and had over 9000 hours, including over 3000 hours in jets, many of the women had comparable hours in their careers. I stopped counting my hours when I reached 10,000 as pilot in command, but had more than 5000 hours when I took the physical. Several of the other women had flown comparable or greater numbers. Jerrie Cobb, the definitive authority on our facts, told me that the average for the Mercury Women was in excess of 4500 hours.

I remain sure that Jerrie Cobb's success at Pensacola was a wake-up call to NASA. The NASA leadership must have believed we would fail the physicals and certainly thought, as Glenn did, that woman pilots with only civilian experience could not fly jet aircraft. But it has been known ever since jets were developed that flying them is only a matter of proper training. Today, if only men could fly jets, the aviation industry would only find one half of the public willing to buy them. Instead, the industry is delighted to sell women corporate jets as well as the new personal jets. Look at the ever-larger numbers of women pilots flying for the airlines and in the military to understand how fundamentally flawed Glenn's testimony was. His assumption that we could not have learned to fly the jets or gained the expertise to handle the Mercury Capsule was, and is, ludicrous. Additionally, a strong case could be made for the fact that the few women who have been allowed to pilot the space shuttles meet or exceed Glenn's stated requirement of being better qualified than the Mercury Seven were, rather than equally so.

The WAFS stepped into fighter planes with only a few hours orientation because of their extensive civilian experience. The WASPs were not required to have more than a few hundred hours before starting their training. They were trained to fly WWII combat aircraft in the same time as the men were trained and flew every aircraft used in combat with a safety record equal to or better than their male counterparts. In order to express such incredible ignorance of women pilots' past achievements, Glenn had to dismiss out of hand the statistical evidence of the WAFS and the WASP's experience as well as Jackie Cochran's ongoing, remarkable achievements in flying jets farther, faster and higher. He testified, without so much as a tip of the hat to the WAFS and WASPs, that such experience would not train a woman for the kinds of stressful pilotage faced every day in test piloting and in the Mercury Space Program.

I've never flown an airplane that could tell the gender of the pilot and doubt that anyone ever will. It is obvious to me that the remaining Mercury Women, each of whom have flown so many thousands of hours, could be trained by any competent instructor to fly the jet aircraft which were the basis of the Mercury Astronauts' military experience. There is no substitute for experience. Jerrie Cobb and Jackie Cochran proved the point. With good flight instruction, any one of us could have demonstrated the competence Glenn made so much of in his testimony.

I find it even more appalling that his testimony before the Subcommittee was unchallenged by any other male in the room when he stated so clearly the real criteria for exclusion. All we had to do was prove that we were better than the men already involved in order to be received with "open arms"? It shouldn't take a feminist to identify the level of discrimination implicit in the widely shared belief that women had to prove superiority before being gifted with the simple opportunity to compete, whether in aviation, space, or any other field. Just as much prejudice is found in Glenn's assumption that none of us could hope to reach the same level of competence he and his fellow astronauts had achieved.

After the Mercury Program had given way to larger vehicles, it was widely reported that Glenn had suggested there were ninety pounds allocated on board for "recreational equipment" that could be filled by a woman. The suggestion flows naturally from the arrogance and condescension of his earlier testimony before the Congress. It clearly represents his and most of the other astronauts' reaction to the scandalous notion of women being given an opportunity to prove themselves in space. This conclusion is confirmed by Gordon Cooper's remarks about brains and dames in space that I quoted previously from Jackie Cochran's book.

The astronauts' reactions beg several questions. How could they have such a narrow understanding of the physical and mental ability of their fellow humans? How could such he-men be so threatened by women and scientists?

I repeat, all the men had done at this point was to ride into space. After all is said and done, it was the scientists who figured out how to get them there, and back, and several of NASA's scientists were women.

Glenn's most recent comments on NASA's failure to include women in the Mercury program repeat that the real reason women were excluded was the "social order" of the time. How convenient for him to forget his own testimony and beliefs which, when presented to the Congressional Subcommittee, were instrumental in building a firewall against the Mercury women.

Jackie Cochran

Sadly, Glenn's testimony before the committee relating to the negative impact of including women in the Mercury Program was essentially supported by Jacqueline Cochran, the only American woman with considerable jet time. In 1962 Jackie faced the prospects of aging and failing health. These factors eliminated her as a candidate for the astronaut corps. She clearly made a choice to support NASA rather than the Mercury Women she had helped create.

What an opportunity was lost! Her own outstanding achievements as a pilot included international speed records. She had flown the fastest jets of the time at Mach 1 and Mach 2. She was uniquely qualified to tell the committee that women could fly the jets quite well, if they were given the chance. But Jackie had worked with the Air Force for many years while leading the WASPs and had relied upon the US military for support in many of her other pet projects. She knew that NASA and the military leadership were opposed to women as military pilots or as partners in space and must have come to the conclusion she should not challenge their position.

Her testimony was damning to the Mercury Women when she proposed additional testing with large numbers of candidates to prove that women were physically capable of becoming astronauts before allowing them into the space program. This proposal made no sense to any of the Mercury Women because we had already passed the tests and had proved ourselves physically. The Mercury Program was limited in numbers of proposed flights and the testing she proposed would make it impossible for women to be a part of that program. When she testified that inclusion of women would unnecessarily impede the Mercury Program, she corroborated NASA's position even though no experts were called to testify how the program would be impeded in any way and no one suggested any problems inclusion of the women might cause. Her support of NASA gave the committee an excuse to support the status quo in the Mercury

Program while giving lip service to some kind of ill defined, parallel program for women to be established in the future. The possibility of an American woman being the first woman in space died in that hearing.

Having had the opportunity to get to know Jackie Cochran while President of the 99s and, later as President of IWASM, I was personally stunned at the report of her testimony. I knew her as a fighter for women in aviation and had great admiration and respect for her. Her failure to stand tall before the committee seemed a betrayal of her own values, let alone the legitimate expectations of the Mercury Women.

Ten years after the hearing, Jackie and her husband Floyd Odlum visited Bob and me while they were on a motor coach journey around the country. It was a remarkable experience for us. Floyd was world renowned as the chairman and majority owner of the Atlas Corporation. He was the highest paid salaried individual in America during the year leading up to the war and his companies built many of American's WWII warplanes. Floyd was already spending most of his time in a wheelchair and Jackie was so supportive and solicitous that the great love and mutual respect between them was obvious. He and Bob enjoyed recalling their own early days in the legal profession while Jackie and I spent a lot of time together. We enjoyed and joined the conversations of our husbands or we talked about aviation in general.

The subject of her testimony before the Congressional Subcommittee came up briefly. She said NASA had led her to believe that any change in the Mercury Program to include women could have cut off Congressional funding for that Program. Jackie knew we would be a success, but didn't support us as the lesser of two evils. She was clearly uncomfortable about talking about her testimony so we didn't stay on the topic very long.

I must admit my admiration of her status as one of the preeminent women pilots remains intact. We shared a warm and wonderful three-day visit with them. Watching her loving interaction with my young son, Darryl, and her cooking special potatoes for our dinner from an old recipe she loved, were worth a whole bunch of supporting speeches. I know that, under different times and circumstances, we could have been friends. She was about as complicated a woman as I have ever known. On balance, she did so much more for women in aviation than almost anyone else that I hope we will write off her performance before the committee as an aberration rather than as representative of the real Jackie. I have made peace with her in memory because her lifetime of achievement more than balances any single mistake that occurred so late in her life.

Geriatrics in Space

Senator Glenn was recently, generously, given another ride in space for NASA under the guise of geriatric research. When the opportunity for Jerrie Cobb to fly on one of the shuttle missions was actively promoted, his silence was as damning as his original, knee-jerk opposition to women in space.

Jerrie is a superbly conditioned pilot whose missionary work in South America, since her rejection by NASA, has been a saga of extraordinarily dangerous jungle flying over a period of many years. In the process she has continued to prove herself as a pilot with few peers and as a contributing woman of the world with few equals.

Since Glenn went back into space as a "specialist" involved in researching the effect of aging on space flight, there are no longer any questions that can be raised regarding Jerrie's comparative qualifications to do the same job. His triumphant ride on the shuttle has been heralded in almost the same heroic terms as his original flight. The failure of NASA to allow Jerrie Cobb a similar opportunity after so many years can no longer be brushed aside as the result of the "social order". It can be argued that the most obvious reason for Glenn's failure to support her inclusion in the research is the simple fact that she would have been even bigger news than he was.

If, as Glenn and NASA claim, research into the effect of the aging process on going into space is a matter of national concern, the inclusion of an older, qualified woman in the program would make good sense. If NASA wants to demonstrate its stated commitment to gender equity, the inclusion of Jerrie would make even better sense. If Glenn wants to demonstrate a real commitment to women's participation in the space experience or express regret regarding his role in killing the possibility of a Women in Space Program in 1962, this is his chance to make amends. As a US Senator, a national leader and an astronaut, he can become a strong, public supporter of the inclusion of one of the Mercury Women in a geriatrics research program for space. To date, however, it remains apparent that his fundamental beliefs still do not include allowing women to compete in his personal arena. He is still unwilling to share the national spotlight.

Women can admire Glenn's courage and achievements while wishing his intellect and experience would give him a broader perspective and understanding. Now that the huge contributions by women on every level are generally recognized, it is sad that the prominent former Senator from Ohio remains trapped in his own tunnel vision of the "social order."

Gender Equity

In retrospect, with 20/20 hindsight, it was probably naive for us to believe that logic and facts would sway NASA to the point of changing their cherished beliefs in the superiority of men. As Congressman Fulton suggested, the notion that jet training was necessary to fly a capsule had been pretty well disproved when HAM successfully rode the rocket into space and came down safely. It would finally be put to rest when Valentina Tereshkova, *the Soviet parachutist*, became the first woman in space. With her flight the fact became even clearer that the Mercury Women were fully qualified. We just were not the chosen gender. NASA's decision to cancel our Pensacola jet training after Cobb's success, allowed it to continue the jet qualifications justification for excluding women for the next thirty years.

I was reminded the other day of NASA Administrator James Webb's speech to one of the larger women's organizations after Congress had failed to change the criteria for astronauts to include women. He addressed his distinguished audience with words to the effect that NASA would be happy to send blacks and women into space when it got safer. I don't know if that put us in third class, or if that meant we were so important they couldn't risk losing us. It's rather remarkable how, so many years later, NASA has survived the breaking of the gender and color barriers without falling apart at the seams. So much for the mythology of the "social order"!

Glenn's incredible and unsupportable belief that only men could fly the jets and that women pilots could not make the life or death decisions required with the same speed and ability of the male test pilots was finally put to rest when the military allowed women to compete as jet pilots. Women in the military now fly the jets, fly as test pilots and otherwise demonstrate over and over again what they proved so long ago in WWII. Glenn and NASA were flat-out wrong.

Fortunately, women have come a long way in gaining gender equity. Most people know the importance of desire, motivation and ability in either a man or a woman. Before she gave her life to space with the other members in the Challenger astronaut crew, Judith Resnick said, "Firsts are only the means to the end of full equality, not the end itself." Who was first in space is no longer as important as the fact that women are now accepted for their ability at every level in NASA's space effort. This is tremendously important because our country should never again make the same mistake of overlooking ability because of gender.

We need look no further than the women who have volunteered and been accepted by NASA since our debacle. Among the astronaut candidates who were selected for training as mission specialists in January 1978 were Dr. Anna Fisher, Shannon Lucid, Judith Resnick, Sally Ride, Dr. Margaret Seddon and Kathryn D. Sullivan. They were selected for their many fields of study: chemistry, medicine, crystallographic x-ray, biochemistry, electrical engineering, missile and surface radar, rocket telemetry, astrophysics, radiation therapy, earth sciences, oceanography, mathematics, boundary layers and heat transfer science, re-entry dynamics, lunar lighting, aerospace medicine, aerospace food, astronaut health and safety and teaching. Of these six astronauts, four are also pilots, although they were selected as mission specialists rather than as potential pilots of space craft.

As I am writing this story, six woman astronauts are pilots beside Eileen Collins, the first woman to pilot a space shuttle. Nancy Sherloc has 2600 hours in US Army Aviation and instructs military pilots. The other five have PhDs or MDs. Shannon Lucid is a member of AOPA and has also logged hours in the Russia's MIR space station. Kathryn Sullivan holds both powered and glider ratings. Anna Fisher is a hobby pilot. Janice E. Voss holds both Amelia Earhart and Howard Hughes Fellowships and Margaret Rhea Seddon has over one hundred hours in the air. By the time this book goes to press, there will be more women shuttle pilots on the astronaut roster.

Eileen Collins was the first woman accepted to train as a pilot of the Shuttle. I am extraordinarily pleased that she has invited the Mercury women to each of her lift-offs. The first time Eileen blasted into space as pilot of STS-63, in 1995, it took my breath away, but left me ready for more. On July 22, 1999, it was even more thrilling to witness her lift off as Commander of STS-93.

I recently read an internal report about Rocketdyne, one of the parts suppliers for Eileen's flight, from a source I consider to be reliable. According to this report, STS-93 was the most dangerous shuttle mission as far as mechanical problems were concerned, since Challenger. I am proud that a woman commander, Air Force Colonel Eileen Collins, kept her cool and led her crew's handling of each situation to a successful conclusion, so STS-93 returned successfully to earth.

Conclusion

Regardless of my comments on John Glenn's pivotal role in NASA's refusal to include the Mercury Women in America's space program, I believe the accomplishments of the Mercury Seven in space, their personal courage and their willingness to lay their lives on the line for our country, fully justify their status as heroes. Having said that, I would note that all any of the Mercury Women asked for was the opportunity to compete on an equal basis with the men for the right to be a part of America's space program. We knew we had the right stuff, the qualifications and the will to lay our lives on the line alongside the men if given the chance. I confess my personal pride that I was one of that group of twenty-five women who volunteered to become candidates for a possible Woman in Space Program and one of the thirteen Mercury Women who passed the grueling astronaut physical. It is my opinion that History will bear out the strong case against NASA and Glenn as far as not including women is concerned.

As a national hero, Glenn occupied an extraordinary position at the hearing and there is no question in my mind that he could have, single-handedly, convinced the committee that women were worthy to compete as candidates for the space program. It is pleasant to imagine what could have happened if, instead of hiding behind old ideas and deprecating thoughts, he had stood up to the fact that the so-called "social order" was wrong and should be challenged. What an even greater hero he would be if he acknowledged women pilots' capabilities and achievements as demonstrated by so many women during his lifetime. It would take no stretch of his imagination. Remember, he knew all about women flying WWII aircraft and the jets. The costs were high indeed; the killing of America's chance to have an American as the first woman in space, the delay in allowing women to pilot spacecraft and a loss of thirty years of example for our young women. The conclusion of the experience of the Mercury Women could have and should have been a positive one for our

We visit our spot: Jerri, Rhea, Janey, Jerrie, Wally, B, and Sarah.

country. Think of the combination: a woman in space program and women's rights. What a win-win that should have been.

When Commander Eileen Collins selected her crew for STS-93, she responded to a question by stating that she was not interested in filling her crew with women. She was interested in filling her crew with qualified astronauts. This is the way it should be. The chance to compete is all anyone can claim as a right. Selection must always come from performance. Historically, women have proven they can compete with men in almost any situation if given a level playing field. Why should space be different?

Studies have been done that show women are most interested in the peaceful use of space exploration. It will take the best of the best to explore and put to peaceful use the findings from our national ventures into space.

But the challenges are not only in space. Space is only one of the incredible frontiers of science and exploration we face. Our young students must be told their place in such adventures will depend on staying in school and pursuing math and science. They are our future, and we must help ensure that they will have the opportunity to compete on an equal basis, if prepared.

As we travel with such accelerated speed into that exciting future of science and exploration, let's all remember what Amelia Earhart said so beautifully in her poetry. *"Courage is the price that life exacts for granting peace."* Although life will almost certainly continue to require individual acts of courage, I believe it is time to for each of us to be prepared to act to do away with any kind of discrimination. Allowing discrimination against any person diminishes each of us. Let us be ever mindful of the value of each individual and not let old habits or old labels like "social order" get in the way of progress again. The costs have always been, and always will be, too high.

Appendix A

Biographies of Mercury Women: 2000

This biographical material is based on the *LIFE* magazine article of June 28, 1962, the 1979 History of the 99s, Al Hallonquist's website www.mercury13.com , and personal information.

MYRTLE CAGLE—38—Myrtle is listed in *LIFE* as a flight instructor. As you have read in "Tethered Mercury," this means I consider her to be the equivalent of a test pilot, plus. By 1961, Myrtle "K" Cagle had flown 4,300 hours. Now, among other things, she is a commercial pilot, teaches the more technical instrument flying and is a certified aviation mechanic. K's also a licensed Nurse and works with the Civil Air Patrol.

JERRIE COBB—32—Jerrie Cobb was the first woman to be tested at the Lovelace Clinic. She was a skilled air racer and an aircraft company executive. Jerrie was taught to fly at age twelve. With a head start on the other Mercury Women, Jerrie progressed further than the other applicants. "After the [Mercury 13] disappointment, Jerrie scrounged up an old twin engine Aero Commander, and [began to fly] Amazon jungles as a Missionary Pilot. …she held four world records in the twin engine class of aircraft… [and received a] nomination for a Nobel Prize for her efforts in the Amazon."

JAN DIETRICH—36—Jan was the twin from Los Angeles with the most hours under her Belt. She was a flight instructor and flew as a corporate pilot for a construction company. When she was tapped for the Mercury program, she already had over 8000 hours on her ticket.

MARION DIETRICH—36—Also professionally active as a pilot, a charter pilot, Marion, a writer and reporter for the *Oakland* (CA)

Tribune, she earned over 1500 hours on her own flying time. Marion died in 1974.

MARY WALLACE FUNK II—24—Wally is the youngest of the Mercury women. She was a commercial pilot at the time of the tests. With irrepressible enthusiasm, she has become a beloved instructor and a noted speaker for women in aviation. Wally recently participated in a space camp program in Russia.

SARAH LEE GORELICK—29—Ma Bell had become AT&T by the time Sarah went to Albuquerque. An engineer and an air racer she had numerous ratings added to her commercial license by 1961. "She now works for the IRS in Kansas, and still pokes holes in the sky for fun, in a Cessna 172."

JANE HART—41—In the article in *Life* magazine, Jane is listed as a mother and a pilot but, as you now know, this does not begin to tell the tale. Jane holds Whirly-Girl license #23, was one of the founders of the National Organization of Women.

JEAN HIXSON—39—A WASP in WWII, Jean was a graduate of Class 44-6 at Avenger Field, Sweetwater, Texas and flew the B-25 twin-engine bomber as an Engineering Test Pilot. In the fifties, with a new career as an elementary and secondary teacher, Jean started an astronomy class and took her students on field trips to NASA's Lewis Research Center in Cleveland. After the Lovelace tests, Jean worked on flight simulator techniques at Wright Patterson AFB in Dayton, Ohio until she retired from the Air Force Reserves as a bird Colonel. Jean died of cancer at age 62.

IRENE LEVERTON—36—With 9000 hours, Irene was a flying school supervisor at the time of the tests. Irene attempted to join the WASP, when she was 17, by using a fake logbook and older friend's birth certificate. Irene continued to fly after the Mercury 13 testing, and is now working with the Civil Air Patrol.

GENE NORA (STUMBOUGH) JESSON—26—worked in aviation demonstration and sales. After the tests, Gene Nora began working for Beech, where she met her husband and helped introduce the Musketeer by flying to 48 states in 90 days. She has written an aviation column, done flight and ground school instructing, owns an aviation insurance business.

Appendix A: Biographies of Mercury Women: 2000 ❧ 249

BERNICE TRIMBLE STEADMAN—37—When *LIFE* magazine released our names, I was listed as owner of a flying school. I stopped counting my hours at 16,000 hours. Al Hallonquist includes details like I hold the highest FAA license and was elected as President of the 99s in 1968. You know more about me now.

GERALDINE (SLOAN) TRUHILL—33—Jerri is the woman who became my roommate in Albuquerque. She was, and is, a corporate pilot. In just a teaser from Al Hallonquist's website, "Jerri T. was first exposed to flying at the age of 4...When she was 15, she began taking flying lessons after school without the knowledge of her parents. By 1960...She was...one of the county's most experienced pilots... [later flying for] Texas Instruments, Incorporated [when they were] developing the Terrain Following Radar (TFR) and "smart" bombs...Jerri T. also participated in numerous Air Races and has a ton of trophies."

RHEA (HURRLE ALLISON) WOLTMAN—32—At the time of the Mercury tests, Rhea had an aircraft brokerage business. She flew all over the United States, including Alaska, and in Canada and Mexico. An air-racer, she towed and instructed in gliders for Cadets at the Air Force Academy. "Shortly after the testing at Lovelace, Rhea stopped flying professionally. She is currently one of the few Parliamentarians left, and is quite in demand" at high profile meetings around the world.

Biographies of Supporting Cast

MARY E. CLARK—As my most-often partner in the air races, Mary was such a reliable person. She went on to become governor of the North Central Section of the 99s, and served on the International Board.

ELIZABETH HARDING—The Marshall Islands are now the legal focus of this woman.

JANEY HART—see also Jane Hart in the Mercury biographies. As the wife of Senator Phil Hart, Jane is also known as a supporter of family and environmental concerns. The interpretation center at Sleeping Bear National Lakeshore is appropriately named after Phil. The Hart family also created the Phil Hart Scholarship for students attending Lake Superior State College in Sault Sainte Marie, Michigan.

JOAN HRUBEC— Joan worked for a man she met at the 1952 All-Ohio Race. Balas-Collet, is a specialized machine-tool company. Eventually she retired as their production manager. She's raced in many events as pilot-in-command (PIC), winning some, and has acted as co-pilot (SIC) for other women. Along with her various activities as a pilot, Joan served as international secretary for the Ninety-Nines and as administrator for the International Women's Air and Space Museum (IWASM) in both Centerville and Cleveland, Ohio. She also acts as a timer and a judge for the All Women Transcontinental Races, now called the Air Race Classic. Joan has over 1600 hours as pilot-in-command.

JACKIE SCOTT—The former governor of the Northeast Section of the Ninety-Nines, Jackie is an IWASM trustee, serves Leelanau County as a Commissioner for Cherry Capital Airport and raises cherries at Flying Scotts Farm.

B STEADMAN—I am now the president of our family business, TC CabCo. This business involves two of my life long pleasures; meeting interesting people and buying cars.

BOB STEADMAN—My husband still has an active legal practice, but he does not put in as many hours to help direct the laws by which we settle disputes. Bob applies his interests and people skills as manager in helping advance TC CabCo. The business keeps us both involved with our children, and gives us many hours of pleasure, sharing stories of the people we meet when we fill in during their busiest hours or take special passengers to their destination.

DARRYL STEADMAN—Darryl and his wife, Eva, helped create our family taxi service. Darryl was vice-president of the business, with the job of keeping the cars in good running order. With automobiles (mostly Lincoln Towncars) constantly on the road, Darryl's was a full time job. Eva is a vice-president with the duties of dispatch and record keeping.

MICHAEL STEADMAN—The Left Bank Café in the gaslight district of Traverse City on the banks of Boardman River is where you can find our son Mike and his wife Patty, an excellent chocolatier. They serve an eclectic mix of fine foods in their small location (good weather seating on the patio/dock) and enhance this with a large luncheon and special events full-catering service.

CAROL STEADMAN—Carol went on to graduate from Brown University as an investment banker.

ELIZABETH HARDING—The United Nations representative for the Marshall Islands.

BERNIECE (BOWERS) VAILLANCOURT—Bernice raised two successful children, enjoys being a busy retired grandmother. We get together frequently when she is not playing golf in the south.

BOB VAILLANCOURT—Bob is an excellent resource for aviation information. He plays a wicked game of golf.

DENNIS WHIPPLE—Denny teaches in Mesick near Traverse City. I participated in a careers program for his students. With their questions, all students keep me on my toes. An exchange student from Russia was in the audience in 1999. After the program, I appreciated his comments and viewpoint of things that I discussed from an American viewpoint.

Appendix B

Development of Women Astronauts: 1960-1999

Mercury Seven Astronauts introduced   1959   1

Jerrie Cobb passed astronaut tests at Lovelace Clinic   1959   2, 6
and Pensacola. Appointed by James Webb as special
advisor to NASA on programs involving women. Jerrie
is quoted as saying she was "the most unconsulted
consultant in government."

B Steadman tested   April 2, 1961   3

Jane Briggs Hart tested   1961   4

Hearing, U.S. House of Representatives   July 1962   5
Congressional Subcommittee on the Selection of Astronauts

Tereshkova flew 48 orbits   June 1963   6
Russian parachutist, a textile factory worker is shown "primping
for orbit" by U.S. media. (The first woman in space could have
been a pilot with thousands of hours flying experience from the
U.S.A.)

Women actively first recruited as astronauts   July 8, 1976   6

Sally K. Ride—first U.S. woman in space   June 18, 1983   6, 7
Physicist from Stanford U. One of six
women selected from 1251 applicants in a
class of thirty-five astronauts.

Eileen Collins—first woman to pilot a spaceship, STS 63   1995   8
Air Force Lt. Col. Discovery pilot. She told the
Associated Press that "Part of the launch that was
exciting for me was to have so many of the women

pilots...during WWII come to see the launch and some who tried to become astronauts back in the early 1960s.

At the Women In Aviation International Conference in 1995, Collins repeatedly thanked her support team and stated she did not want women alone. "I want the best of the astronauts along with me." Audience applause, of course, indicated they knew some women would be among the astronauts on the flight she would command in the future.

—first woman to command a space shuttle, STS 93   1999   8, 9
"I've admired pilots, astronauts and explorers of all kinds. I also think it is important to point out that I did not get here alone." Then she included women pilots from the barnstorming years through WWII and the Mercury Women. Commander Collins is now a member of the WIA Hall of Fame and a Ninety-Nine. She impresses the hell out of me.

The list of thirteen women listed as part of the Astronaut Corps by the *Chicago Tribune* in 1988 reflects the thirteen original women who passed the Mercury Astronaut Physical at Lovelace Clinic. Six of these astronauts had already been in space. This allows me to dream that if they had made us part of the Corp in 1961, I really might have gone into space. Change that to read, "I really would have flown in space!" BTS

1. *We Seven* by the astronauts themselves
2. www.jerrie-cobb.org
3. Lovelace Clinic Letter to Steadman dated March 24, 1961
4. www.friends-partners.org/~mwade/astro/hartjane.html
5. Newspaper article announcing the hearing.
6. Joyce Dell'Acqua/Smithsonian News Service byline *Chicago Tribune*, October 1988.
7. *Lewis News*, March 1987.
8. Personal notes, e-mail from Mercury Women and various newsclippings, JMC.
9. Note in BTS e-mail showing report describing the problems.

Additional Reading and Addresses

20 hrs. 40 Mins Amelia Earhart.
30th Reunion Slated For Participants in the First Women Astronaut Testing Program Sara Rickman, IWASM Quarterly 1991, Vol. V, Issue 3. p 3.
A Good Time for Vultures Jerri and Joe Truhill
Amelia Earhart's Daughters: The Wild and Glorious Story of American Women Aviators from World War II to the Dawn of the Space Age Leslie Hainsworth and David M Toomey, Harper 2000, 336 pp. The second half of the book is about the Mercury Women.
Arial Vagabond Bessie Owen, Liveright Publishing Co, 1941, 256 pp.
Astronauts: the first twenty-five years of manned space flight. Bill Yenne, 0671081942.
Aviatrix Elinor Smith, Harcout, Brace, Jovanovich 1981, 298 pp.
Barefoot in the Sky: An Autobiography Sheila Scott.
Congress and the Nation 1945-1964
Girls Can't Be Pilots: An Aerobiography Margaret J. Ringenberg with Jane L Roth, 305 pp.
Hidden Heroine-Fay Gillis Wells Sara Rimmerman, 1999, 29pp.
High, Wide and Frightened Louise Thaden, Stackpole Sons, 1938.
I Live to Fly Jacqueline Auriol.
Jackie Cochran: An Autobiography with Maryann B Brinley Bantam 1987.
Ladybirds and Ladybirds II Henry M Holden & Captain Lori Griffith, Blackhawk Publishing 1993, 324 pp. First Lady Astronaut Trainees (Vol. II) p 198-204.
Last Flight Amelia Earhart, Orion Books 1988, 140 pp.

Listen! The Wind Anne Morrow Lindbergh, Harcourt, Brace & Co 1938, 275 pp.
Littlest Soldier, The Nancy Hopkins-Tier
My God! It's A Woman: The Autobiography of Nancy Bird Angus & Robertson 1990, 216 pp.
North to the Orient Anne Morrow Lindbergh, Harcourt, Brace & Co 1935.
Rising Above It An Autobiography Edna Gardner Whyte with Ann Cooper 1991.
The Right Stuff Tom Wolfe, Bantam 1983, 368 pp.
Sisters in the Sky: Vol. I—The WAFS Adela Riek Scharr, The Patrice Press 1986, 531 pp.
Sisters in the Sky: Vol. II—The WASPS Adela Riek Scharr.
Sisters of the Wind: Voices of Early Women Aviators Elizabeth S Bell, Trilogy Books 1994, 198 pp.
Solo Woman: Gaby Kennard's World Flight Gaby Kennard, Bantam Books 1990, 216 pp.
The Sound of Wings: the Life of Amelia Earhart Mary S Lovell, St Martin's Press 1989, 412 pp.
Straight on Till Morning: The Biography of Beryl Markham Mary S Lovell, St Martin Press 1991.
The Sugar Bird Lady Ann Sperry
Those Wonderful Women in Their Flying Machines: The Unknown Heroines of World War II Sally Van Wagenen Keil, Four Directions Press, 1990, 401 pp.
United States Women in Aviation 1919-1929 Kathleen Brooks-Pazmany, Smithsonian Institute. 1991, 57 pp.
United States Women in Aviation 1930-1939 Claudia M Oaks, Smithsonian Institution Press 1991, 70 pp.
United States Women in Aviation1940-1985 Deborah G Douglas, Smithsonian Institution Press 1991, 138 pp. WACOA p64, 89-90, 92.
Voyager Jeanna Yeager and Dick Rutan with Phil Patton, Alfred A Knopf, 1987, 337 pp.
We Charles Lindbergh
We Seven by *the Mercury Astronauts Themselves*, circ. 1960 LIFE.
West With the Night, Beryl Markham, North Point Press 1983, 294 pp.
Women Aloft Valerie Moolman, Time-Life, 1981, 173 pp.
Women and Flight: Portraits of Contemporary Women Pilots Carol Russo, Little, Brown & Co 1997, 192 pp.

Margaret A. Weitekamp, "The Right Stuff, The Wrong Sex: The Science,Politics, and Culture of the First Lady Astronaut Trainees, 1959-1963". Master's dissertation., Cornell University. Advisor: Richard Polenberg.

International Woman's Air & Space Museum, Burke Lakefront Airport, 1501 Marginal Rd, Cleveland, OH 44114
www.iwasm.org

Mercury Women		www.mercury13.com

Ninety-Nines, Inc International Headquarters, Box 965, 7100 Terminal Drive, Oklahoma City, OK 73159		www.ninety-nines.org

Women In Aviation		www.womeninaviation.com (has excellent links)

Women at NASA		www.jsc.nasa.gov
		www.jsc.nasa.gov/Bios/htmlbios

Glossary

Advanced Trainers (ATs)-Complex Aircraft used to teach military pilots.

Aerocommander-An executive high-wing, twin engine aircraft.

Aeronca-Single engine, two place training and personal taildragger airplane. It has a small high wing and is very economical. The outstanding visual characteristic is an extension of the trailing edge of the tail fin and a fuselage that clearly shows the supporting structure of the fabric.

Air Mass-A large block of air with uniform temperature and humidity across a horizontal section.

Airframe & Powerplant (A&P)-Designation of license held by aviation mechanic.

Basic Trainers (BTs)-In military codification, somewhat sophisticated; prop is adjustable but the landing gear is fixed.

Bendix Trophy-First given in 1931, women began to compete in this prestigious coast to coast race in 1935. In 1936, Louise Thaden became the first woman to win this race. She took home both the regular prize and the women's prize.

Cessna-An aircraft company noted for high wing aircraft.

Cherokee-A Piper built low wing single engine airplane for training and personal use.

Chord-Shape of wing; width and depth.

Civil Air Patrol-A para-military organization that helps young people learn how to fly and conducts air search and rescue missions. During WWII, the C.A.P. trained civilians as airplane spotters and performed U.S. coastal anti-submarine patrols.

Commercial License-License designation required by the FAA for pilots who fly for hire. Earned by demonstrated performance of skill (a minimum of 200 hours flight time), an extensive written and flight check.

Copilot-A pilot who assists the pilot-in-command.

Cross-country flight-Any flight that originates at one airport and lands at least one other airport more than 50 nautical miles from the departure airport. Students going after increasingly higher licenses and ratings are required to make cross-country practice flights and landings of increasingly complex leg flights and landings.

Dead Reckoning-A form of navigation that includes predetermination of track across the ground using forecast wind information to compute aircraft headings, ground speed and time to destination.

DME-Distance Measuring Equipment; measures distance between radio source and Airplane, another dial on modern flight panel.

Dove-An executive airplane built by de Haviland; Twin engine, seating for six-eight.

Drag-In aviation, a retarding force that is the opposite of thrust.

E6B-A hand held mechanical computer used to determine the effect of measurable factors (wind direction and speed, heading, etc) to predict flight information (fuel consumption, time to destination, etc).

Enroute Pilot Report (PIREP)-Pilots radio weather conditions they are experiencing to control towers, radar centers and flight services. This networking creates a refined source of information for pilots who are considering a flight through the same air mass.

Ercoupe-Single engine airplane, twin-fin tail. A conventional steering wheel moves ailerons and rudder simultaneously, so it only has two controls. Ercoupe introduced tricycle gear to the private pilot.

Fairchild PT-19, PT-23-Low wing, tandem cockpit, fixed-gear, trail-dragger airplanes used for primary training.

Fixed-base Operator (FBO)-The owner of a general aviation servicing business. Also the location of the same. FBO personnel do everything they are allowed to help a pilot. They may sell aviation gasoline, oil, charts, fuses. They do or can arrange for repairs, pump up airplane tires, call ahead on request, run to get your lunch, loan a pilot their car, provide a lounge in which a pilot can rest, sleep, shower, offer popcorn, pop and snack machines, telephones and a weather computer. The best ones are air-conditioned with chairs and couches to sink into and offer tooth brushes and other forgettable products. This is where the line boy works. They may also offer instruction and aircraft rental and sales.

Flight Line-At an airport, this is the area where active aircraft are tied down.

Flight Plan-Filed with the control tower, it lists required items such as time to destination, alternative airports and passengers. If a pilot does not close the plan within expected time, it triggers a search.

Flight Service-Pilots contact Flight Service for weather information.

GI Bill-These acts of congress fund education of military personnel leaving the service.

Gold Seal Certificate-Designates a more experienced and well-qualified flight instructor with special privileges. They can give flight tests.

Groomed Airplane-An airplane that has been cleaned up for a race. All components of the plane are tightened, anything that will cause avoidable drag will be removed or closed.

Ground Effect-Compressed air that forms between a wing and the ground that creates an unusual amount of lift within one wing span of the ground. It is a kick in the pants! Especially in a race.

Ground Loop-When an airplane spins around in a small circle on the ground. The airplane is out of control and is dangerous to both plane and people.

Ground-speed check-A formula using time in the air and distance covered to predict the time of a continuing flight.

Human Counter-An early MRI; used for medical diagnosis.

Instrument Flight Rules (IFR)-FAA rules a pilot must follow if weather conditions reduce visibility and cloud clearances. These rules create spacing between airplanes. IFR flight is regulated by radar control centers around the world.

J-3 Cub-A slow, stable, tandem-seat, high-wing, taildragger airplane with exposed cylinder heads and a flat spot on the tail fin. Fabric covered, the fuselage shows wood that shapes it. These antiques are often painted a bright yellow.

Lear-An aircraft company that developed the first general aviation autopilot. They are famous for their sleek executive jet aircraft.

Leg-A flight from one airport to another in a flight that lands at more than one airport.

Lift-Force created by air passing over the wing, allows an object to rise into the air.

Light-gun-A light signal used by controllers at an airport. The use of red, white or green filters in a certain pattern sends message to airplane without radio communication.

Luscombe-Single engine, two-seater with tapered wing, metal fuselage, taildragger.

Mae Wests-Life jackets that are flat on the chest until inflated with compressed air. They are so named because the person wearing one looks well endowed.

NAFEC-National Aviation Facility, Experimental Center.

NASA-National Air and Space Administration.

Ninety-Nines-International organization of women pilots founded by ninety-nine pioneer women pilots. Amelia Earhart was the first president of the organization.

OSHA-Occupational Safety and Health Administration.

Pilot-in-Command-The pilot who is responsible for decisions made for an aircraft.

Piper-An aircraft company that builds training and executive airplanes.

Pitot tube-A tube that projects from the leading edge of a wing that provides pitot (ram air) pressure to the airspeed indicator.

Primary Trainers (PT)-See basic trainer (BT)

Private Pilot License-In the year 2000 requires an instructor signature for a student with 40 hours of flying; 20 or more with a flight instructor, 10 or more solo hours, 5 hours of solo cross country (one of at least 150 nautical miles with three landing points and one leg of at least 50 nautical miles), and 3 hours of night flying. Earning the license can take longer.

Pylon-Points in the landscape that airplanes race around. In a formal race these are tall, bright cones on the ground.

Second-in-Command-A pilot who would take over as requested by the Pilot-in-command or in an emergency when the PIC is unable to act.

Spirit of St Louis-The name of the plane Charles Lindbergh designed and flew as the first pilot to cross the Atlantic Ocean.

Stinson Reliant-Massive braced gull wing, taildragger, with a cowled radial engine. The unique chord of the wing is not straight. People enter the plane by climbing a short ladder that does not retract.

Taylorcraft D, T'craft-A single engine, side-by-side two-seat airplane with a high wing. The wing was long and wide, convex on both surfaces, without any dihedral. The large, upright tail fin has a distinct flat spot on the rudder. The fuselage is tapered and smooth, the cylinder heads are exposed. Fabric covered, taildragger.

Tetrahedron-Used at some airports to indicate the direction of the wind. The small end points in the direction from which the wind is coming..

Thompson Trophy-Given to the winner of a speed race around pylons held in the U.S. Coveted internationally.

Tie down-To attach an airplane to the ground with a series of ropes. An unattended airplane can fly without power in the right wind conditions.

TravelAir-Very small, tapered low wing, twin engines, vertical tail fin, dihedral. It required a complex, multiengine rating to fly it.

Vertigo-A condition of the inner ear with a loss of equilibrium. A sensation of dizziness and losing one's balance. In an airplane, vertigo makes a pilot feel as if the airplane is tumbling.

Visual Flight Rules (VFR)-FAA rules a pilot uses when the weather is within the established boundaries.

VOR, Very high frequency Omnidirectional Range station-A ground-based electronic navigational aid transmitting very high frequency navigational signals, 360 degrees in azimuth, oriented from magnetic north.

WACOA-Womens' Advisory Committee On Aviation

WAFS-Women's Auxiliary Ferry Service

WASP-Women Airforce Service Pilots

Whirly Girls-Women Helicopter Pilots' Organization

X-craft-Experimental aircraft first used to break the sound barrier. Used for new designs until the FAA has approved them.

Resources: Bernice Trimble Steadman, Jody M Clark, Mike Stock of the Northwestern Michigan College Flight Department, the *Oxford American Dictionary, The Proficient Pilot* by Barry Schiff, *AOPA's Handbook for pilots, Chronicle of Aviation* by Chronicle Communications, and *A Field Guide for Airplanes* by M R Montgomery and Gerald Foster.

Bibliography

Chapter 1-25: Steadman, Bernice Trimble. Taped Interviews and discussions by Jody M. Clark, April 1993-September 2000.

Chapter 2: Vaillancourt, Berniece (Bowers). Taped Interview by Jody M. Clark, 1993.

Chapter 3 and 11: Vaillancourt, Berniece and Robert. Taped Interview by Jody M. Clark, 1993.

Chapter 5: *The Timetables of History: A Horizontal Linkage of People and Events* by Bernard Grun, Touchstone Edition, Simon and Schuster 1982, 591 pp.

The Peoples Chronology: A Year by Year Record of Human Events from Prehistory to the Present by James Trager, Henry Holt 1992, 1126 pp.

Chapter 10: Hrubec, Joan. Interview by Jody M. Clark. May 26-27, 1994. Response from Joan Hrubec 1995.

Scott, Jackie. Interview by Jody M. Clark. February 8, 1998.

Chapter 15: Hart, Jane (Briggs). Taped Interview by Jody M. Clark, September 8, 1993.

Hart, Jane (Briggs). Observed interviews, British journalist. September 25, 1997.

Steadman, Bernice Trimble. Observed interviews, British journalist. September 25, 1997.

Hart, Jane (Briggs). Transcribed interviews by Margaret A. Weitekamp. October 7, 1997.

Steadman, Bernice Trimble. Transcribed interviews by Margaret A. Weitekamp. October 7-8, 1997.

Ninety-Nines AWTAR Program Booklet 1957.

Chapter 16: Promotional material, *Harebell*. Travel Journal by Jane Howard. Taped Interview with Jane Briggs Hart, September 8, 1993.

Chapter 17: *The Timetables* and *The People's Chronology.*
   Cochran, Jacqueline Papers. Letters. Dwight D. Eisenhower Presidential Library.
   Cochran, Jacqueline Papers. Letters. Jackie Cochran Archives, from Jane Anderson.
   Lovelace, W. Randolph II, M.D. Letters. Dwight D. Eisenhower Presidential Library.
Chapter 18: Taped Interview with Jane Briggs Hart, September 8, 1993. Observed interviews September 25, 1997, October 7, 1997.
   Steadman, Bernice Trimble. Letters. Personal Archives.
   Steadman, Bernice Trimble. Lovelace Clinic Graph. International Women's Air and Space Museum.
   The Delt-Air 250: A Tribute to Herb Dean EAA Sport Aviation, February 1962
Chapter 19: Cochran, Jacqueline Papers. Letters. Dwight D. Eisenhower Presidential Library.
   Cochran, Jacqueline Papers. Letters. Jackie Cochran Archives, from Jane Anderson.
   Lovelace, W. Randolph II, M.D. Letters. Dwight D. Eisenhower Presidential Library.
Chapter 20: Steadman, Bernice Trimble. News clippings. Personal Archives.
Chapter 21: Steadman, Bernice Trimble. Letters. Personal Archives.
   Steadman, Bernice Trimble. WACOA documents. Personal Archives.
Chapter 22: Steadman, Bernice Trimble. News clippings. Personal Archives.
   Ninety-Nines' History 1997.
Chapter 23: Taped interviews with Steadman family 1995.
   Visit with Berniece Vaillancourt 1995.
Chapter 24: Taped interviews with Steadman family 1995.
   Visit with Berniece Vaillancourt 1995.
Chapter 25: *Lewis News*, 29 March 1987. "Early Testing of Women Pilots Showed They Could Be Astronauts."
Chapter 26: *Amelia Earhart's Daughters: The Wild and Glorious Story of American Women Aviators from World War II to the Dawn of the Space Age* Leslie Hainsworth and David M Toomey, Harper 2000, 336 pp.
   The Lovelace Center: Pioneering in American Health Care Jake W. Spidle, Jr., University of New Mexico Press 1987, 209pp.

Women In Aviation and the subsequent *Women For Aviation* magazines, and in reports given at the annual Women In Aviation International Conferences.
www.jsc.nasa.gov/Bios/htmlbios/collins.html. Eileen Marie Collins (Colonel, USAF) Biographical Data. Lyndon B. Johnson Space Center.

Index

A
A. C. Spark Plug, xv, 4, 39, 77
Aerobatics, 19, 47-49
Air Force Cadets, 22
Aircraft Owners and Pilot's Association (AOPA) 177-178, 243
Airshows, 18, 43
Airway Underwriters, 203
Albuquerque, New Mexico, 128, 133, 143, 145, 156, 161, 232-233, 248-249
All-Women's International Air Race (AWIAR), 65-76, 166-171
All-Women's Transcontinental Air Race (AWTAR), 58-60, 111-124, 180-188
Alpena, Michigan, 18, 33
Amelia Earhart's Daughters, 235, 254, 263
Anfuso, Congressman Victor, 162
Angel Derby (*also see* AWIAR), 66
Ann Arbor, Michigan, xiv, 61, 102, 200, 205, 211-215
Army Air Corps, 7, 10, 37, 38, 107
Arnold, Pat, 113
AuSable River, 105
Auerbach, Julie, 202
Aviation Trail, 226

B
B Steadman & Co., xiv, 215
Basic Trainers (BTs), 40, 257
Batista, 71, 129
Baty [Chabrian], Dr. Peggy, 228
Beech, Olive Ann, 220
Bellou, Eftbynia, 225
Bendix Trophy, 36, 257
Bera, Fran, 36, 58, 73, 79, 183, 185
Biofeedback, 215
Bishop Airport, 13, 22, 33, 50, 66, 84, 95, 197
Book-Cadillac Hotel, 58
Bosco, Caro (Bailey), 47
Bowers, Berniece (*also see* Vaillancourt), xiii-xiv, xvii, 9, 13-29, 33, 48, 57-58, 83, 85-88, 129, 213-214, 251, 263
Bowman, Bob, 55
Boy Scouts 204
Boy Scouts of America, 204
Brunswick Corporation, 112, 168
Bunker, Zaddie, 73-74
Burke Lakefront Airport, 228, 256
Bush Pilots, 8

C
Cagle, Myrtle, 247
Canada, 37, 58, 75, 142, 166, 189, 195, 249
Capital Airlines, 94
Caribbean, xiv, 138-141, 143, 146, 148-149
Carmein, Amy, xviii, 227
Carpenter, Liz, 172, 236
Casey, Lew, 220-221
Centerville, Ohio, 221, 223-224, 229, 230, 250

Chamberlain, Clarence, 3, 61
Chicago, Illinois, 30, 97, 195, 210, 218, 253
Chonoski, Bob, 47, 53
Christie Family, 14, 189-190
Christie, Laura (*also see* Whipple), v, 190-191
Civil Air Patrol (C.A.P.), 18-20, 22, 220, 257
Clark, Mary E., 61, 77, 111, 113, 117, 120, 166-170, 183, 185-188, 197-198, 249
Cleveland, Ohio, 130, 228-229, 248, 250, 256
Cobb, Jerrie 130-131, 156, 158, 161-163, 178, 231-234, 236-238, 241, 245, 247, 252
Cochran, Jacqueline (Jackie), 38, 58, 130-133, 157-165, 220-221, 234, 236, 239-242, 254, 263
Collins, Eileen, 164, 230, 243-245, 252
Columbine, 98, 124
Congressional Record, 235
Congressional Subcommittee on Astronaut Selection, 162, 234, 239-240, 252
Cooper, Gordon, 163, 238
Crane, Margaret, 36, 58, 77
Crawford, John, 201
Crowley Boiler Works, 57, 111, 168
Culinary Institute of America, 206
Cuple, Bob, 96
Curtis Condor, 3

D

Dayton, Ohio, 69-70, 208, 221-223, 226-228, 248
de Bernardi, Fiorenza, 225
Dean, Herb, 153, 263
Detroit City Airport, 86-87
Dietrich twins, 156, 247
Dow, Alden B., 222-223

E

E6B, 118, 258

Earhart, Amelia, 5, 32, 56, 62, 178, 214, 220, 224, 230, 234, 243, 246, 254-255, 260
Embry-Riddle Aeronautical University, 55, 228
Ercoupe, 27, 85-86, 91-92, 258
Experimental Aircraft Association (EAA), 87, 153

F

Fede Caproni, Marie, 225
Federal Aviation Authority (FAA), 66, 97-98, 102, 154, 158, 172-173, 175-178, 202, 220, 228, 249, 257, 259, 261
Federation Des Pilotes, 224
Fixed-Base Operator (FBO), 27, 36, 80, 258
Flickenger, Brigadier General Donald, 130-131, 231
Flint Aeronautical, xiv, 50, 53-54, 78, 94, 95
Flint Journal, 42, 121, 154, 193
Flint, Michigan, xv, 3-4, 18, 24, 27, 30, 32, 36, 39, 42, 47, 51, 53, 55-59, 66, 77-78, 83, 86-87, 91, 94, 98, 102, 108, 114, 119, 129-130, 151-154, 180, 191, 193, 196, 208, 210, 214-215
Francis Aviation, xiv, 6, 8, 19, 38, 40, 43, 46, 50, 98
Francis, Jerry, 39, 104
Frasca, Rudy, 100
Friendship, xv, xvii-xviii, 4, 13, 15, 17, 19, 57, 61, 75, 87, 111, 145, 153, 193, 218, 234
Friendship Seven, 234
Fruehauf, Ann, 54, 79
Funk II, Mary Wallace (Wally), 245, 248

G

General Motors Air Transport, 26, 83-88
Geyer, Howard, 115
Gilles, Betty, 178

Gillespie, E.W. (Turk), 51
Gillis Wells, Fay, 224, 229
Glenn, John, 149, 162-163, 231, 234-239, 241-242, 244
Glennan, Admiral T. Keith, 231
Glidden, Charles (Charlie), 55, 91
Gold Seal Certificate, 175, 259
Gorelick [Ratley], Sarah Lee, 248
Gotsch, Kathy, 230
Greece, 224-225
Gunsights, 53

H

Halaby, Najeeb, 173
Hammond, Alice, 58, 172, 178
Harbor Springs, Michigan, 77-79, 86
Harding, Elizabeth, 197, 249
Harebell, 138-143
Harmon Trophy, 38
Hart, Janey B., xiii, xv, xvii, 29, 74, 121-128, 133, 137-138, 141, 143, 145, 148-153, 157, 161-163, 168, 172-179, 196, 234, 245, 248, 249
Hart, Senator Phil, 125-126, 128, 161, 174, 249
Hath, Dale, 50, 54, 78, 94
Helms, Dorothy, 216
Hendrickson [Leslie], Amelia, 56
Higgins, Leah, 77
Heintz, Shirley, 223
Hitler youth, 139
Hixson, Jean, 248
Hoffmann, Hermann, 226
Hopkins-Tier, Nancy, 220, 224-226, 229, 255
Howard, Jane, 138, 141, 262
Hrubec, Joan, xiii, 65-76, 112, 121, 228, 230, 250, 262
Hurrle [Allison Woltman], Rhea, 249
Hutton, Betty, 112-113
Hyde, Louise, 138, 141, 178

I

International Women's Air and Space Museum (IWASM), 158, 163, 219, 220-229, 240, 250

J

J-3 Cub, 84, 259
Jacobson, Leonard, 14-15
Jaguar, 88-89
Jessen, Gene Nora (Stumbough), 148-149, 178, 248
Jobst, Vern, 62-63
Johnson, President Lyndon B., 162, 172, 264
Joo, Jacqueline, 178, 208

K

Kennedy, Rose, 124
Kenyon, Teddy, 5, 53
Key West, Florida, 69, 71, 73
Kloze Klozet, The, 209-210
Kolp, Jimmie, 178, 220
Krakel, Dean, 220

L

Leigh, Josset, 142
Leverton, Irene, 248
Link trainers, 50
Lindbergh, Anne (Morrow), 5, 63
Lindbergh, Charles, 62-63, 234, 255, 260
Los Alamos, New Mexico, 137, 146, 149
Love, Nancy (Harkness), 37
Lovelace Clinic 133, 146, 150, 152, 247, 252, 253, 263
Lovelace II, PhD, Randolph W. (Randy), 130-133, 148, 155, 157-161, 233
Luhta, Connie, 229
Lyman, Mott, 91

M

Maloney, Betty, 138
Mapes, Ann, 216

Mason, Sammy, 48
McDonald, Margie, 230
McKaye, Sammy, 80
Mercury Seven, 130, 232, 236-237, 244, 252
Mercury Women Biographies, 247-251
Michigan Aeroclub, 58
Millane, Bob, 43-45, 61
Miller, Betty, 79, 178
Mott Foundation, 91

N

Nagle, Nadine, 230
National Air and Space Administration (NASA), xvii, 131, 148, 153, 155, 158, 161-164, 231-244, 248, 252, 256, 260
Ninety-Nines, Inc. (99s), 56-66, 73, 78, 87, 112, 121, 123, 162, 178, 181, 186, 202-203, 219-221, 223-224, 229, 250, 253, 256, 260, 262-263
  Michigan Chapter, 56-57, 202
  Ohio Chapter, 65
Northwestern Michigan College (NMC), 32, 261
Noyes, Blanche, 5, 58, 66, 71, 178, 220

O

Odlum, Floyd, 131, 157, 241
Omlie, Phoebe, 5, 61

P

Parker, Jackie, 230
Pedacopter, 225
Pele, Mary Claire, 224-225
Pensacola, Florida, 131, 157-158, 161-162, 232-233, 236-237, 242, 252
Powder Puff Derby, *see* AWTAR
Prochazka, Gertrude, 42, 58
Pylon racing, 83-84, 180

Q

Quamby, Lucille, 79-80, 112-113, 168

R

Ramsey, Jean, 6, 106
Renninger, Doris, 220
Resnick, Judith, 243
Reynolds, Jean P., 57
Rickman, Sarah, 229
Ride, Sally, 162, 164, 227, 243
Rocketdyne Corporation, 243
Rose, Ralph (Uncle Bulgie), 7-10, 23-24, 32, 38-40, 51, 92
Royce, Margaret, 210
Rosco Turner 36
Ruth, Babe, 79, 123

S

Sarat, Sarah, 204
Sault Sainte Marie, Michigan (the Soo), 14, 75, 189-190, 196, 249
Schneider, Dr. Richard, 212
Schulhoff Lau, Susan, 163, 223
Scott, Doris, 222-223
Scott, Jackie, 75-76, 250
Seacrest, Dr., 136, 148
Selfridge Air Force Base, 94
Shamburger, Page, 178, 219-220
Sloan [Truhill], Geraldine (Jerri), 145, 157, 233, 249, 254
Smith, Eloise, 79
Snyder, Gary and Carol, 227
Southern Michigan All Ladies Lark (SMALL Race), 77-80, 121, 201
Spirit of Saint Louis, 35, 62-64
Steadman family, 189-207
Steadman, Bob, xiii, xiv, 99, 191, 194, 250
Steadman, Carol, 197, 251
Steadman, Darryl, v, xiv, 199, 200-206, 214, 240, 250
Steadman, Michael, v, xiv, 204, 207, 250
Strother, Dora (Dougherty), 79, 178

Stubbs, Colonel George P., 95
Sub-committee on the Selection of Astronauts, *see* Congressional Subcomittee

T

Taylorcraft (T'craft), 7-8, 11, 16, 18, 23-24, 26, 32, 33, 36, 84, 85, 92, 260
Tenth Air Force Reserve, 94
Tereshkova, Valentina, 242, 252
Thompson Trophy, 36, 228, 260
Traverse City CabCo., 250
Traverse City, Michigan (TVC), 32-33, 74, 76, 79, 121, 200, 203-204, 206-208, 210, 212, 214-216, 250-251
Trimble Aviation, 33, 92-110, 168, 175, 183
   Open house, 100
Trimble Family, 189-190
Trimble, Laura (*also see* Whipple)
Truhill, Jerri (*also see* Sloan)
Turner, Roscoe, 36
Twining, Bob, 102

U

University Hospital, 156, 211-212

V

Vaillancourt, Berniece (Bowers), xiii-xiv, xvii, 9, 13-22, 24-27, 33, 48, 57-58, 83, 85-88, 129, 213-215, 251, 262-263
Vaillancourt, Bob, xiii, xvii, 21-29, 83-87, 91-92, 106, 251, 262
Vietnam 128
Volunteers In Service To America (VISTA), xiv, 216
von Braun, Dr. Wernher, 151-153
Von Iberg, Datelieb and Joanne, 139

W

Walton, Burr, 6-8
Washington D.C., 66-68, 70, 84, 122, 124-126, 172, 174, 183, 186, 196, 221, 234
Webb, James, 161, 242, 252
Wetherill, Helen, 77
Whipple, Dennis (Denny), xiii 197, 251
Whipple, Laura Christie Trimble, v, 4-5, 12-13, 15, 190
Whipple, Raymond (Ray), 3, 5, 13, 91, 191
Whipple, Richard (Dick), v, 3, 13, 15, 191
Whyte, Edna Gardner, 79, 121, 169, 255
Wiles, Ivan, 86
Wilson, Lois, 59
Wolf, Connie, 177-178
Women in Aviation, Inc., 227-228
Women's Advisory Committee on Aviation, (WACOA), xiv, 172-173, 175-179, 208, 255, 261, 263
Women's Airforce Service Pilots (WASP), 38, 71, 132, 220, 223, 236-239, 248, 255, 261
Women's Auxiliary Ferry Service (WAFS), 37-38, 236-238, 255, 261
World War II, 11, 30, 254-255, 263
Wright, Asahel, 223, 226, 229
Wright-Patterson Air Force Base, 208, 214, 230

X

X-craft Program, 155

Y

Young, Dottie, 60, 178

Z

Zonta International, 62-63, 210, 215-216

TETHERED MERCURY

*Text design by Mary Jo Zazueta in AG Garamond
Text stock is 60 lb. Vellum
Printed and bound by Data Reproductions,
Auburn Hills, Michigan
Production Manager: Mareesa Orth*